"Where is everybody?"

The question became known as Fermi's Paradox. In 1950 the physicist Enrico Fermi asked this question while discussing the possibility of extraterrestrial civilizations. People reasoned that an advanced civilization able to travel at only a fifth the speed of light would be able to colonize the galaxy in a million years. Our Milky Way Galaxy is more than 13,000 million years old. On Earth, after its surface cooled from molten lava to solid rock, only a tiny part of its present age was needed for the first life to appear. And there are billions of planets in the Galaxy where a similar story could play out—time and place enough for many vast civilizations to have come and gone—and some to exist still.

Fermi asked, mightn't one expect to see some sign of alien civilizations—stray radio signals, hints of Dyson Swarms around other stars, and indeed, why not passing travelers? But nothing . . . It has been called the Great Silence.

As it happens, timing is everything. When the Great Silence finally ended—when the question of the existence of extraterrestrial intelligence was finally dispelled with the first contact from an alien civilization of staggering advancement and age, humanity faced the most extraordinary surprises in all its history. This account is one story of those surprises and challenges that began for Earth's billions and four friends in particular on New Year's Day 2052—*anno Nuntii*—the year of the message.

AFTER

THE

SILENCE

AFTER THE SILENCE

DAVID LOSCHKE

AdLonge
To Reach Far...

AFTER THE SILENCE

Copyright © 2022 by David Loschke

All rights reserved. No part of this publication may be reproduced, distributed, or transmitted in any form or by any means, including photocopying, recording, or other electronic or mechanical methods, without the prior written permission of the publisher, except in the case of brief quotations in critical reviews and certain noncommercial uses permitted by copyright law. For permission requests, write to the publisher addressed "Attention: Permissions" at publisher's email address below.

ISBN: 978-1-7378745-0-8 (Hardcover with Dust Jacket)
ISBN: 978-1-7378745-3-9 (Hardcover Case Laminate)
ISBN: 978-1-7378745-1-5 (Paperback)
ISBN: 978-1-7378745-2-2 (eBook)

Publisher's Cataloging-in-Publication Data

Names: Loschke, David Carl, 1945– author.
Title: After the Silence / David Loschke.
Description: Kettle Falls, WA: Ad Longe Publishing LLC, 2022.
Identifiers: LCCN: 2022930053 | ISBN: 978-1-7378745-0-8 (hardcover) | 978-1-7378745-1-5 (paperback) | 978-1-7378745-2-2 (eBook)

Subjects: LCSH Human-alien encounters--Fiction. | Extraterrestrial beings--Fiction. | Genetic engineering--Fiction. | Science fiction. | BISAC FICTION / Science Fiction / Alien Contact | FICTION / Science Fiction / Genetic Engineering | FICTION / Science Fiction / Hard Science Fiction

Classification: LCC PS3612 .O772 A48 2022 | DDC 813.6--dc23

Any references to historical events, real people, or real places are used fictitiously. Names, characters, and events are products of the author's imagination.

Book interior and cover design: 1106 Design

Publisher: Ad Longe Publishing, Kettle Falls, Washington
Publisher's email address: inquiries@adlongepublishing.com

TO JOYCE, JON & GARY

CONTENTS

Chapter 1 – The Message	1
Chapter 2 – Upheaval	11
Chapter 3 – Dark Thoughts	23
Chapter 4 – The Visitors	33
Chapter 5 – Second Revelation	45
Chapter 6 – Reaction	61
Chapter 7 – Waiting	71
Chapter 8 – More To Ponder	83
Chapter 9 – Unexpected	95
Chapter 10 – Decisions	107
Chapter 11 – History Lesson	119
Chapter 12 – Seeking Consensus	131
Chapter 13 – Invitation	143
Chapter 14 – More Waiting	149
Chapter 15 – Journey	163
Chapter 16 – Anticipation	177
Chapter 17 – Arrival	191
Chapter 18 – New Moon	201
Chapter 19 – Getting Acquainted	213
Chapter 20 – New Neighbors	225
Chapter 21 – Chats with Sandra and Grace	239

Chapter 22 – Chats with Gerry and Walter	247
Chapter 23 – Chats with Ellen and Nia	259
Chapter 24 – Chats with Jim and George	267
Chapter 25 – Conspiracy	275
Chapter 26 – The Incident	291
Chapter 27 – Aftermath	307
Chapter 28 – The Gift	313
Chapter 29 – Farewell	327
Epilogue	337
About the Author	339

"Two things fill the mind with ever new and increasing admiration and awe the oftener and more steadily we reflect on them: the starry heavens above me and the moral law within me."

—Immanuel Kant
in *Critique of Practical Reason*
and on his gravestone

Chapter 1

THE MESSAGE

Canberra, Australia, Monday, 1 January 2052

"Jim, Jim, wake up ... Jim!" Ellen shook him. "Jim!" Her voice was urgent. "JIM, wake up NOW!" She shook him again.

Jim's mind swam out of a dream. He opened his eyes and squinted from the bright sunlight pushing through the curtains. "What's wrong?" he croaked. "Oh my head aches. What time is it?"

"It's already eleven in the morning," she said. "Hurry! Get up and listen to the news!"

"Is there coffee?" he said while pulling on his track pants.

"Never mind that! Listen to the news! I just switched it on." She turned and shouted through the open bedroom door, "TV louder." Jim followed her out of the bedroom and sank onto the sofa. He sighed and closed his eyes.

"JIM, wake up ... please! Pay attention!" Ellen said firmly. He forced his eyes open and saw a special breaking news announcement on Australia's ABC network. The newsreader's face was somber. Jim's sluggish mind began to focus on his words.

"... have been reported from dozens of countries first in the eastern hemisphere and working westward early this morning. Government reactions are not yet clear. Beijing and Moscow have remained quiet so far. EU sources, London, and Washington are urging calm and have suggested the likelihood of an elaborate hoax. The prime minister in Canberra has called an emergency meeting of his cabinet this morning, and they are still in session. We have had reports that all Australian military commands have been raised to a heightened alert status, but government sources have refused to confirm the reports."

Jim's head swirled. He looked at Ellen and said, "What? Military put on alert? What's happening?" She looked at him, her face filled with worry. They turned back to the screen, Jim fully awake now—coffee forgotten.

The TV announcer continued. "It has just passed midnight in London, and New Year's crowds around the city have been subdued. The message began to be received this afternoon across Eastern Europe, but it had already been passed westward to European governments and the US government by military and embassy outposts in East Asia and the Pacific. The message has now been linked into digital networks around the world, but we are informed that it is still being broadcast from a stationary source overhead—a signal that appears to be coming from space. It is now 7:10 p.m. in New York City, and plans for the New Year's Eve celebration in Times Square are being questioned. New York City's mayor has said they will go ahead and that this entire story must be a hoax. US President Kaitland has called for calm, but anonymous sources say that she is examining the present DEFCON status of US military forces. Stay tuned for further announcements. We will continue with this breaking story throughout the day on this station."

The TV audio changed to unobtrusive classical music and the screen showed a view of Australia from space with the overlaid message—Stand By.

Ellen gripped Jim's hand. They looked at each other again. "Did you hear anything else before I woke up?" Jim asked struggling to put it all together. "I mean, what's this message they keep talking about?"

2

"I don't know," Ellen said. "I had just gotten up and started to listen when I ran to wake you. It's not even the New Year in the United States yet. What do they mean about canceling celebrations and putting the military on alert?"

"This is big," Jim muttered and put his arm around her. Ellen was strong and did not frighten easily, but he could see the alarm in her eyes. He felt it too. "I don't get it," he said. "How did all this happen last night? When did it—"

The TV announcer began to speak again. "We have just been given further information about the radio signal that is now being received across the Western Hemisphere. Our reporters stationed in Washington, DC have learned that the message was first detected a little more than twelve hours ago at US bases in Alaska, Hawaii, and Wake Island. Ordinary televisions tuned to any of a half-dozen of the old UHF bands still operating in parts of Micronesia also received it. Radio astronomy facilities in the region picked it up too on their dedicated bands but at reduced power. We have confirmed now that it was first noticed on the east coast of Australia at approximately 11:00 p.m. last night."

"What!" exclaimed Jim. "This means it was already happening while we were watching the fireworks on the bridge last night!"

The TV announcer continued. "The Canberra Deep Space Communication Complex and the Australia Telescope Compact Array facility have analyzed records with cooperating observatories and report that the signal began within seconds of midnight, Universal Coordinated Time, directly above the International Date Line's primary longitude. The signal source has not moved and has continued to come from the same point in the sky while Earth rotates beneath it. We have just learned that the message is being received simultaneously in eleven of Earth's most widely spoken languages."

Jim felt a tingling sensation in his legs and moving up his spine—his mind was racing. Ellen had begun to tremble. Softly, almost to himself, Jim said, "The signal coming from space in eleven of Earth's languages? Who could be doing this?"

The phone rang. Both Jim and Ellen jumped. Jim put it to his ear and heard the familiar voice of his friend Gerry on the line. "Jim? Jim—have you been listening to the news? It's a hell of a thing! I've been trying to call you for ten minutes, but the phones have been jammed."

"Gerry! I'm so glad it's you! Have you heard anything about this message they're talking about? Where is—"

"Yeah, just found it! Go to channel 372—it's being repeated there in English. You won't believe—" The phone call dropped out.

Ellen shouted "What did he say? What did he—"

Jim shouted at the TV, "Channel 372!"

The TV displayed a text in English scrolling down the screen while a gentle voice was speaking. They could not be sure whether it was a man's voice or a woman's. They were still trying to make sense of the few words they had heard when the voice and text stopped.

The last words still displayed on the screen said, "We hope someday to become your good friends. This message will repeat in ten of your minutes." Below that sentence was a large number "10:00" that began ticking down: 9:59, 9:58, 9:57 …

Now Jim was trembling. "What does that mean—ten of *our* minutes? Can this be what it seems to be?"

The phone rang again. Again they both jumped. Jim picked it up. Gerry's voice said, "Jim, the phones are still acting up. Did you find the message?"

"Gerry! Great, you're back! Yes, we found the channel, but the message was just ending when we tuned in, and now we need to wait for it to repeat—it said in ten minutes."

"That's OK," said Gerry. "It's been repeating for hours—I guess since it first started. It doesn't take that long to play through."

"Put it on speaker Jim!" Ellen shouted.

He touched the button.

"But what's going on?" Jim said too loudly. "Have you heard the whole message?"

THE MESSAGE

"Yes, we have," Gerry said. "If this is real, everything has changed. And I mean *everything—really*! Jim, Ellen—they say they've come from another star and want to be our friends!"

Silence.

"Jim, Ellen—are you still there?"

Jim, still trembling said, "Can this be a hoax? Are you kidding us?"

Gerry answered, "Hey, not me! And I think not anyone. If what we've heard so far about the signal is true, it would be incredibly hard to pull off. If the signal is stationary in the sky, no country on Earth could feasibly do it. I'll go to the space complex soon to see if I can find more information about the signal, but from reports, it must be strong. Jesus! They claim they're way outside the Kuiper Belt! Even if the signal is tightly beamed, they must have a lot of power behind it. I'm sure the signal direction is being checked right now to see if it's moving across the background star field. If it's still, then it's not coming from any conventional Earth satellite. And it can't be coming from one of the Moon bases either—the Moon's in the wrong position now. Same thing goes for the small Mars base. Besides, Mars wouldn't have that kind of power." He sounded maniacal in his excitement. This was Gerry's biggest possible secret dream coming true. "Hang on!" he said. "The message is about to start. I want to listen again too. Will call back later." The line went dead.

Jim and Ellen turned to the TV.

The screen clock was ticking down the last few seconds to zero. The gentle voice returned, and the text of its words began to scroll on the screen.

"Hello people of Earth. We have come to your planetary system in peace and friendship and are sending you this message to introduce ourselves. Our home world is in another star's planetary system far from here. We are a group of explorers and scientists who have been traveling in this sector only to learn. We know that this message will be very surprising to

5

AFTER THE SILENCE

you. Please do not be alarmed. We are still far away from you. We have entered a region you call the Oort Cloud, but we have parked in your planetary plane in an outer orbit around your star. Our position is far outside the zone you call the Kuiper Belt. We hope you will agree that this is a respectful distance for visitors to wait while we become acquainted.

"We have been able to learn much about you and a number of your languages from your many radio transmissions. We are eager to learn more, and we expect that you also would like to learn about us. You can send us questions on these carrier frequencies—569, 581, and 593 MHz. We will monitor these frequencies to collect your questions, and we will attempt to answer first those asked most often. You may begin transmitting to us at any time. We will be able to receive any signals transmitted at your normal power levels.

"Be aware that we are 8.9 light-days distant from your Sun. Therefore, with time to consider your questions and time for signal travel in both directions, you will need to wait at least eighteen days for our responses.

"Most important, please do remember our assurance that you are in no danger from us—we intend you no harm. We hope someday to become your good friends. This message will repeat in ten of your minutes."

As the message ended, Jim and Ellen hugged each other silently. Both were stunned. Jim's mind drifted back to the night before—New Year's Eve, Sydney Harbour, the fireworks on the bridge, the happy crowds. They had arrived in Canberra only three days earlier to begin a long overdue vacation. Gerry, Jim's old friend who lived in Canberra with his wife Sandra, had invited them to spend New Year's Eve with them for a beautiful summer night of watching the spectacular New Year's celebration around the Sydney Harbour bridge. After the celebration finished past midnight and the four made the long drive back to Canberra, they were tired but happy. It was such an adventure, and the world seemed so fresh and carefree that summer morning of New Year's Day 2052. But now ... now nothing felt normal. Nothing felt the same. And nothing felt carefree.

While Jim and Ellen had a late breakfast in their sunny kitchen, they distracted themselves with watching the gorgeous parrots on the large bird feeder just outside the kitchen window.

"Just look at the dazzling blues and reds on those crimson rosellas!" Jim said.

"I know," Ellen said. "And the two rainbow lorikeets out there too—I looked them up in the bird book. Such brilliant orange, blue, and green. And they act so cute!"

The parrots went on with their miracle of poses and postures unaware of their admirers and of their own beauty and charm.

Jim said, "We had such plans for this special trip, our first holiday in seven years! I wanted everything to go so well. Now, I don't know ..." His voice trailed off.

The next hours passed in a haze of repeated news announcements. A break from the repetition came with a discussion between an astronomer and an expert linguist, both from the Australian National University in Canberra. First the astronomer, Professor Yang, began explaining just how far away 8.9 light-days actually is.

She said, "I realize many of your listeners won't be familiar with this terminology. But it's used in astronomy because the distances across space are so vast. Light and radio signals, which are the same thing but with different wavelengths, travel at 300,000 kilometers per second. Just think of that for a moment! Earth is about 150 million kilometers from the Sun, and it takes a little more than eight minutes for light to travel from the Sun to Earth. Now think of where the aliens say they are—where it takes 8.9 days for light from the Sun to reach them. They are way *way* farther out than our outermost planets like

Uranus and Neptune. They're approximately 1,540 times farther from the Sun than Earth is. The Sun looks like a bright star from where the aliens say they are parked."

The program moderator said, "Thank you, Professor Yang. That brings the distance home to me much better and I'm sure for our listeners too. Now I would like to ask our linguist, Professor Adama, what he thinks of the fact that the message is being broadcast in eleven of Earth's major languages."

Professor Adama cleared his throat and said, "I was astonished when I first heard that. In the past, I have heard plenty of discussions that if we ever were to make radio contact with intelligent extraterrestrials, we would first need to go through a long difficult period of trying to establish some kind of common means of communication like say beginning with a demonstration that we understand mathematics. You may have heard examples such as sending a series of pulsed signals representing numbers and then showing how numbers can be added, multiplied, and so on. It would be a very long process before either side could advance to communicating broader concepts that are common in language. And yet now we hear their first message broadcast in eleven of our languages, and all are spoken and written flawlessly. Their pronunciation and grammatical sentence construction are essentially perfect. This would be difficult even for a group from Earth to accomplish using the same voice. It's why I think the message is some kind of spectacular hoax. Yes, I heard the message say that they learned much about us from our radio transmissions, but still I find it all very implausible."

"Thank you, Professor Adama," the moderator said. "I'm sure many of our listeners would agree with you and hope that you are right as well. So back to you, Professor Yang. Does your expertise shed any light on that point?"

Professor Yang replied, "Yes, I know it's hard to believe that a first contact could come to us this way. But remember that Earth has been broadcasting radio transmissions at significant power levels for

150 years. That means that our radio signals have spread out 150 light-years in all directions past many other stars. If there is anyone out there within that growing sphere of our signals, they could have been listening to us for a long time. Let's suppose that in addition to the ability to travel between stars, these aliens possess far advanced computing powers compared to ours—say quantum computers that we're still working to perfect. Let's further suppose that these aliens might be more intelligent than we are. It's possible that decoding our languages might be child's play for them.

"So far we know nothing other than the fact that the radio signal is coming from a stationary source in space. That simple fact alone renders a hoax exceedingly implausible if not impossible for anyone from Earth. The Moon and Mars bases are both in the wrong position and so are ruled out. Anything else someone from Earth could broadcast from space would need to be sent from a satellite or maybe some secret rocket probe. All conventional versions of those would be moving across the background star field. Perhaps a special rocket might be possible that is already far out in the area of the outer planets and on an outward-bound trajectory so perfectly aligned that it appears stationary to us against the background stars for a time. But I can't imagine anyone from Earth expending such extraordinary cost and effort as that for the sake of a hoax. I'm keeping my mind open."

Throughout the afternoon, Jim and Ellen tried calling Gerry and Sandra. Jim was anxious to talk more with Gerry who was an astrophysicist at the Australian National University. But the phone network had deteriorated even more, and messaging didn't work either. Only once were they able to have a brief conversation with Sandra, who said, "Gerry went to the space complex to see if he can find out anything new about the signal. He has contacts with colleagues at other universities—they have special communication links that might still be working."

"That would be great if he can find out anything at all," Jim said. "We're getting nothing new over the news channels now."

"We'll be sure to call you," Sandra replied. "I don't know how late he plans to stay there. I've tried to phone him but have had no luck getting through. I guess I'll need to wait until he comes home. But we'll try to phone you then."

They said their goodbyes, and that was the last phone communication Jim and Ellen had that day. By that night, news organizations had reached to the bottom of the barrel and were interviewing a number of people who passionately described how they had been kidnapped by aliens in the past.

Ellen said, "I don't need any more of this. TV off."

They both were behind with their sleep from the night before, and by bedtime Ellen was barely able to navigate to the bedroom. Jim was surprised he didn't feel sleepy yet. His mind refused to let go of the incredible day they had just finished. His thoughts kept roaming through the news reports, especially the words of the message. *What if this is real? Everything would be different!*

Caught in a liminal space bounded by *what was* and *what is yet to come*—he went to bed, his mind spinning in a whirlpool of both excitement and anxiety.

Chapter 2

UPHEAVAL

Washington, DC, Monday, 1 January 2052, 7:00 a.m.

On the other side of the world in a cold gray winter's dawn, US President Rachel Kaitland was preparing for a meeting of her cabinet in an effort to consolidate what was known about this unprecedented event. She had first been made aware of the message almost twenty-three hours earlier when it was forwarded from US military bases in the central Pacific. The previous day had been full of chaotic conferences with the Joint Chiefs of Staff, intelligence experts and congressional leaders—all with no concrete results.

President Kaitland had set the meeting for 0800. She insisted on scheduling all White House meetings formatted in military time, a carryover from the first two decades of her career in the US Space Force. Rachel Kaitland was only the second woman elected President of the United States. She—like gold nuggets caught in the crook of a mountain stream—was a treasure of glittering talents caught in a single person. Her stellar intelligence enabled her to achieve a PhD in particle physics by the age of twenty-one. Matching her intellect, her social skills guided her through brilliant accomplishments in the still new Space Force, and her emotional intelligence attracted allies from all directions. By the age of forty-two, her restless ambition lured her into a new way to make her impact on the world. She ran for a US Senate seat in her home state of Massachusetts and won despite the predictions of scornful pundits. After another brilliant five years in the Senate, her fame and popularity propelled her into the presidential race of 2048. Her win in November of that year, a story in itself, was almost ensured by her sheer charisma and the recognition at last in the electorate of the need for science in the service of government. Although some were envious of her meteoric fame, most considered her an authentic star who deserved her success.

The clock in the Cabinet Room of the West Wing showed 0801 when President Kaitland called her cabinet meeting to order. This was no ordinary New Year's Day morning. Though few in the room had slept at all the previous night, none had been kept awake by partying. Members of support staff sat on chairs around the room's perimeter, and everyone wore sober expressions, their eyes on the president.

She began, "I won't dwell on preliminaries here. All of us have listened to the message. It purports to be from extraterrestrials, but that point should not be considered settled until we have more evidence. For now, evidence does support that the message broadcast originates from beyond the Kuiper Belt. Our intelligence agencies are still examining every possibility. If it proves to be a hoax, we already have measures in place to deal with it. But if in time we can only conclude that an extraterrestrial presence in our solar system is real, we must consider this issue in new ways. It is possible that everyone on Earth is in grave danger. This morning, I want us to discuss the situation as if the aliens are, indeed, real. First I want to hear views from the Department of Defense and the Department of Homeland Security. Secretary Wainwright, would you care to begin?" She turned to the Secretary of Defense, a seasoned veteran in his late sixties.

"Thank you, Madam President," Oscar Wainwright said as he gathered his thoughts. "I met this morning with the Joint Chiefs of Staff, and we have agreed on a few preliminary assumptions beginning with the obvious. If the alien message is genuine, we must consider whether they are deceiving us about their true intentions and whether we might be facing a threat. We have no way of knowing the answer and must, therefore, consider whether we have effective means of defense if needed. At present our technology cannot project force to where we believe the aliens are located. The key outer planets that NASA used before with slingshot velocity boosting are not in position to be helpful. Even if they were, the journey for one of our current rockets would require more than four hundred years to reach the alien's position. We have no way of taking a fight to them. However if they are planning aggressive actions, we can guess that they would send attack vessels closer to Earth. In that case, it is possible that we could mount an effective defense by modifying existing rocket technologies and using nuclear warheads from our existing defense installations. Since no existing armed missiles are designed for attack or defense operations even as high as low Earth orbits, those steps would require substantial preparation time.

"But this level of analysis is naive in a sense. Based on the aliens' use of our languages, we can assume they have been watching us for a considerable time and know a good deal about us. They might plan threats less overt than outright military attacks. Because we do not know anything about their capabilities other than an assumed technology more advanced than ours, different threat possibilities include biological warfare, chemical attack, extreme environmental disruption, disruption of communication systems, electrical grids or other vital infrastructure, undermining civil order through misinformation campaigns, or even some means of affecting human psychology in ways that we have never encountered. These few examples do not exhaust the possibilities that can be imagined.

"The first concrete step we believe to be of most value is to harden all government communications systems. We need to assume that the aliens can intercept anything broadcast or even passing through internet server systems. Most government systems are protected with encryption technology, but there are a few gaps. I have already asked my department's own experts to begin an exhaustive analysis of current system weaknesses and strategies for improving all digital security. I have also asked our intelligence experts to devise false communications to send as 'bait'—a kind of test to see whether the aliens reveal they have become aware of them. If they are as far away as they say, we'll have no possible responses for a long while, and they might be too clever to rise to such bait. But it's an easy stratagem to try. Meanwhile the DoD is continuing to analyze this problem with all of our resources. Thank you, Madam President." Secretary Wainwright sat down.

President Kaitland heard nothing that surprised her. She had already thought through all of these points as, indeed, had most defense security analysts. She had hoped for a surprising new insight from one of the defense teams but was disappointed. "Thank you, Secretary Wainwright. Let's next hear from the Department of Homeland Security. Secretary Taylor?"

Mason Taylor nodded and stood. "Thank you, Madam President. I, too, have met with our subdepartment heads along with our best departmental intelligence analysts. Secretary Wainwright has well expressed the lines of discussion we also pursued, and I won't repeat his analysis. In agreement with his last point, my department's responsibility for cybersecurity is the most critical issue, and we will aim to collaborate with the Department of Defense. I have set Homeland Security experts to work on strategies to harden our cybersecurity. If we find a way to mount any kind of effective defense against the aliens, we must be able to develop and refine our plans in secret. My chief of staff is coordinating a summary of a DHS plan to have for you by 1800 today."

Taylor was succinct; something Kaitland appreciated in a colleague. "Thank you, Secretary Taylor" she said. "Since the full cabinet is in attendance, I ask now for anyone else to share ideas for us to consider."

There was silence around the table except for shuffling papers and creaking chairs. Normally such an invitation would call forth any number of people who enjoyed hearing themselves talk, but not this time. She interpreted the silence as an unmistakable sign of just how much the world had changed. She thanked everyone, adjourned the meeting and left the room with her chief of staff. She thought, *This is a New Year's Day like none before.*

Meetings like this one were taking place in national capitals all around the planet. A few European countries like Britain and France and others like Russia, China, and India discussed the adequacies of their own defenses that were formidable by Earth's standards. But they arrived at the same grim conclusions as the Americans. Other countries with lesser capabilities were only able to discuss ways to maintain civil order and preserve their status as independent nations in case there should be a collapse of world order. No one knew what might come next.

The first four days of January passed before phone circuits became almost dependable again. By the end of the second week of January, the things one might expect to happen were in progress. World stock markets had plummeted. Runs were in progress at certain banks as a result of careless rumors. Prices for gold, other precious metals, strategic resources and food futures were sharply up. People were stockpiling food, and buying fewer consumer goods except for emergency and communication equipment. Travel plans were canceled. China and Russia closed their borders, and half of the European countries were threatening to do so. In the US, the Canadian border was open, but the Mexican border was closed. Most US churches were filled to overflowing on the first and second Sundays of January. And the internet was sprouting three times as many scams as usual. The same scenarios played out in southern Asia, in South America, and across Africa—similar to events during the two worldwide coronavirus pandemics of 2020 and 2025.

As to the transmission of questions to the aliens, governments moved to take control. Most democratic governments required citizens to submit their questions to a government agency for vetting and only then allowed transmission by government facilities. Some repressive governments attempted to block all nongovernment questions on the nominated frequencies but failed because many of their citizens had enough technical knowledge and older equipment that could be modified and used secretly. A few governments like North Korea kept quiet about their activities. Independent nongovernment groups operated in most nations with each promoting its own agenda and transmitting its own specific questions to the aliens whether in an approved way or not. Meanwhile, media outlets broadcast endless discussions over which questions were most important to ask and provided endless lists of questions already submitted.

Prominent personalities were generous with their own idiosyncratic views. One, for example, on the first day after the message was received urged everyone never to use the word *aliens* for fear of offending them. She suggested that all should use the term *visitors* until we might learn their preferred name. Another celebrity urged all nations to unify their armed forces as soon as possible under one command and begin training for coordinated defensive actions. Various tabloid-style media supplied a copious stream of suggestions from unknown contributors that could most charitably be described as hilarious, uninformed, or deranged. All news sources were saturated with talk of the aliens.

Canberra, Monday, 15 January

Jim remembered one time as a boy when he had driven a shovel into a large ant mound and opened it. The whole area seemed to explode with what looked like millions of ants racing in all directions. He thought that if it were possible to look down on humanity from space and see massed human reaction to the message, it might look a lot like the ants.

The aliens' message continued to broadcast nonstop twenty-four seven. Tracking radio telescopes were able to link together to confirm that the broadcast source was in the planetary plane and, indeed, was positioned directly over the International Date Line at midnight there on 1 January. Jim, with his lifelong hobby of astronomy, understood all of this right away, but Gerry could not be restrained from explaining it in excruciating detail much to Ellen's and Sandra's amusement. They could still be amused between moments when they were deeply worried. Although all four of them feared what yet might come, they were sometimes like friends entering a chapel for a funeral and for a moment forgetting themselves to laugh at some private joke before disapproving glances snapped them back to their surroundings.

"It's wild! It's amazing," Gerry said. "That far from the Sun, their orbital period is a little more than 60,400 years! From here they look like they're standing still. Heck, they're 52 times farther out than Neptune, our most distant planet. The big European Southern Observatory in Chile linked up with other radio interferometry facilities and measured the signal source's movement over the last couple of weeks. Its movement across the background stars is so slow that it's taken time to get a reliable measurement. But they showed it's just the right amount of movement to confirm the distance the aliens are claiming."

In his excitement, Gerry went on before anyone else could get a word in. "The aliens are also at a point in their orbit where they will have a clear view of us for the next six months before Earth swings around later to move behind the Sun and interfere with radio contact. This seems like more than a coincidence, and most figure that they planned their timing carefully. Man, how I would love to know what they're up to!"

During those first weeks of January, Jim and Ellen had grown closer to Gerry and Sandra. Jim and Gerry were already old friends.

They first met five years earlier when Gerry came to the University of Arizona to work with another astrophysicist analyzing backlogged data from the Webb Space Telescope. It was a short three-week visit, and Sandra who was a medical doctor needed to remain in Canberra with her duties at the hospital.

Jim, a professor of evolutionary genetics at the University of Arizona in Tucson, first saw Gerry when he gave an evening guest talk sponsored by the university's Department of Astronomy. In Jim's student days, he had studied mathematics and physics before becoming fascinated by genetics. But he had been an astronomy nut all through his adolescence and had never given up his fascination with physics. He often made time for lectures like Gerry's.

Gerry was a tall, red-haired, lanky fellow with a cheeky grin and an engaging Australian accent who radiated cheerful competence. Jim, not as tall nor as athletic looking, took a liking to him right away. After the lecture, Jim made his way to Gerry and asked him what he thought about a new idea published about dark matter. Gerry was happy to talk, and they soon discovered they had many of the same interests including a similar puckish sense of humor.

In those few weeks they began a lasting friendship. Jim and Ellen hosted Gerry to dinner several times at their small house in the Catalina Foothills just north of Tucson. They got him hooked on American-style Mexican food too at local restaurants and took him on a couple of hikes around the Tucson area to show him the beauties of the Sonoran Desert. One day as a special treat for Gerry, they drove out to Kitt Peak to tour the old observatory facilities there and take in the spectacular views of the surrounding desert and mountains. After Gerry returned to Canberra, they stayed in touch often with video chats that allowed Gerry's wife Sandra to become part of the group as well. On those calls, Ellen and Sandra found that they too had many views in common, and genuine friendship soon grew between them.

All were transfixed by what had happened since New Year's Eve and by what might happen next. For all four, their lives had been transformed in ways particular to each but also in ways that demanded contact and reassurance from each other. Without noticing, they became a new family spending most evenings together talking about the day's events or watching the news broadcasts. And so these four people, improbably bonded together, found themselves facing the single most significant event in all of recorded human history.

One evening after dinner, relaxation demanded a break from talking about the aliens. Jim wondered aloud how simple life "seemed" when they were younger.

"Well of course," Sandra said. "You know how looking back ten years adds a little magic to your memory making things seem happier or more poignant than anything in the present. When I'm honest, I remember that medical school was exhausting and sometimes almost overwhelmed me. My mother was a doctor and warned me it would be like that. Yet now, those midnights in the anatomy lab with the smell of formaldehyde and a cadaver as my only company or the all-night study binges before big exams—now I remember them as happy times. How crazy is that?"

"I know what you mean!" Ellen said. "When I was trying to cram everything possible into my head to get ready for my life-or-death dissertation defense, you would think I would remember it as an awful time, but no. It's an exciting memory, somehow even a happy one."

"That's because you passed with flying colors," Jim laughed. "And your faculty advisors were singing your praises around the department." Jim held up his glass. "I raise a toast to my lovely wife and her brilliant successes!"

When the laughter died down, Ellen turned to Sandra and said, "Remember on New Year's Eve—you were just about to tell me how you and Gerry met when the fireworks started. How about finishing the story?"

Gerry interrupted before Sandra could start and said, "Well hey, that story will show how important it was for me that Sandra did become a doctor."

Sandra laughed and said, "Looks like he'll be telling the story."

"Sorry, honey," Gerry chuckled, "but OK—I will. Well, Sandra happened to be the attending physician in the emergency room one Sunday afternoon eight years ago when I arrived with a broken leg. I had been rock climbing in Nemadgi National Park and had a bad fall. My friends wrapped up my bleeding leg in a towel and rushed me in. One of the broken bones was sticking out through my skin, and I was a mess. But never mind; when Sandra came over and unwrapped the bloody towel, I was dazzled—her bright eyes, the timbre of her voice ... All I could think about was, is she already married? By the time she put me back together and gave me a smile like sunshine, I was ready to propose." Sandra was blushing and reached over to give his hand a warm squeeze.

"And did she say yes immediately?" Ellen asked laughing.

"Ha—I'm not crazy!" Gerry answered. "I waited and bided my time. It was a serious campaign—I was smooth. All I did that afternoon was memorize her name tag and persuade her to give me her phone number." More laughter.

"Really smooth!" Ellen chuckled.

Gerry continued unruffled. "I can tell you that before the cast came off, we were dating seriously."

Sandra laughed and said, "Well, he was Gerry's version of smooth. But I liked it, and anyway he had caught my eye too that afternoon. Not many like him ever show up in the emergency room." Then she turned to Ellen and said, "But what about the two of you?"

Like Gerry, Jim jumped in and said, "I want to tell this one. Maybe it's not quite as exciting as a medical emergency, but it was just as 'out of the blue.' Ten years ago when I was at the University of Washington, I spotted an evening talk in the Philosophy Department that was about the same subject I study but from a philosophical viewpoint. It turned out that Ellen was giving the talk, and I had the same reaction that Gerry just described. Hey, what can I say—I'm a guy! It took a lot of effort to pay attention to what she was saying, but I did get enough to be impressed! By the time she was finishing, I couldn't think of anything other than I wanted to see her left hand better."

Sandra stopped her laughing enough to ask Ellen, "What was your talk about? I'm not convinced that Jim caught that much of it."

Ellen laughed and said, "I happened to be a visiting fellow there but in the Philosophy Department. My specialty is the study of systems of ethics and morality. That's what I talked about—my view of what factors can bring about change in ethical systems. It's been my passion, and I love that Jim understands that and understands how important it is. Jim and I are in different disciplines, but we're kindred spirits. When we both secured appointments at the University of Arizona, it was a dream come true. And now everything is up in the air again, I can't stop wondering how this new reality will change our lives."

Jim observed, "Yes ... we had a nice little break just now, but we can't seem to stay away from the aliens for long."

Chapter 3

DARK THOUGHTS

Canberra, Friday, 19 January

Now in the pandemonium that had erupted all over the planet, Jim found it difficult to think about anything but the arrival of the aliens—the first proof ever not only of life beyond Earth, but also of intelligent life no less! All he wanted to do was think about the implications of this now very different world and follow breaking news every day. How did these new beings come to be—how would they view us—how similar to us might they be—what might we learn from them? And, of course, the riveting question underlying all those—were they a danger to humanity? Was it only a ruse when they said they wanted to be our friends? Would the typical patterns of the natural world he knew so well as a biologist prevail?

Friday evening Gerry stopped by after work to have dinner with them. Sandra had to work the evening shift at the hospital and couldn't come. After dinner they settled down with cups of tea, and Gerry drifted into talk about theories of alien civilizations. Ellen was already overtired and before long excused herself to have an early night.

After the goodnights, Gerry fixed his eyes on Jim and said, "You know how serious this is, don't you?"

Jim groaned and said, "Sure I do! Since I was a teenager crazy about astronomy and science fiction stories, I've thought about the Fermi Paradox—Enrico Fermi's famous question in 1950—where is everybody? Why haven't we found evidence of intelligent extraterrestrial life before? The Drake Equation that came later estimated the probability of intelligent life elsewhere in the galaxy but was criticized as only speculation. Since then though, new evidence has accumulated that supports some of the equation's steps more and more. It's still the clearest way to think about the problem. And now ... now we have an answer! But my gosh, we have even more questions! What else will happen?"

"Exactly!" Gerry said. "Before New Year's Day, hardly anyone ever thought about it—most people don't have a grasp of our galaxy's scale or the number of places intelligent life might appear. It's always been a mystery why we haven't picked up any sign before ... well, until now!

"The galaxy is more than thirteen billion years old. We're just new kids on the block. It's possible that vast civilizations could have come and gone thousands of millions of years before we appeared, and plenty of others could be out there right now. The immensity of space has kept many of us thinking that we couldn't be the only ones. And of course now ..." His voice trailed off.

"I know!" Jim said. "And even that immensity can be conquered by an advanced civilization over enough time. If alien spacefarers could travel at a fifth the speed of light, they could spread across much of the galaxy in less than a million years. It's not unreasonable to think we might spot stray radio signals or see signs of things like a Dyson swarm—vast collections of orbiting alien structures that partly block the light of a star. But we haven't. It's been called the 'Great Silence.' It begs for an explanation. Of course I also know the theories about why we haven't detected any signs. That's what you're worried about, isn't it?"

"Sure," Gerry said. "That's definitely what I'm worried about—ideas like the 'Great Filter,' the 'Dark Forest Hypothesis,' the 'Transcension Hypothesis,' and a handful of others."

"That's right," Jim said. "There are lots of ways in the Great Filter concept of how an intelligent civilization could destroy itself like through war or by mismanaging its own environment. If the civilization goes extinct before reaching a highly advanced stage, we wouldn't hear from them. But it's the Dark Forest idea that really grabs me. It falls right out of my field of evolutionary biology—the struggle of different life forms for existence and the dangers they face competing with each other. If our understanding is right, all intelligent life would have evolved out of that struggle and would see that the same thing could happen at other places in the galaxy too. We can assume that each civilization would see its own survival as its primary interest. We can assume that each would grow and expand outward. Each would recognize that other civilizations would do the same and would see the others as competitors for space and resources. Since each grew out of the struggles and dangers on its home planet, each would consider others a possible danger. Interstellar distances are so great and communication so difficult that trust would have no easy opportunity to develop between civilizations. Instead, webs of suspicion and fear would prevail. Other civilizations out there might be much stronger and perhaps malevolent. A wise civilization would try not to be noticed once it reached such a conclusion. It would cloak itself in silence by any means its technology could provide. It would guess that naïve young civilizations that thoughtlessly sent their radio communications into space might be found and destroyed. It's dangerous out there. It's like a dark forest where each animal does its best to avoid being noticed because it never knows what peril lurks behind the next tree."

"Yes," Gerry said. "I find that logic persuasive too. And we're playing a very dangerous game ourselves. For 150 years we've been broadcasting our soap operas, sitcoms, and sports events out to the whole universe without a thought. The signals are a bubble three hundred light-years in diameter now and expanding outward at the speed of light. It's a beacon saying here we are—come find us. And now someone has!"

"I know, and I'm worried," Jim said. "There are even more frightening ideas too. I'm sure you've heard of von Neumann Machines."

"Sure! You must be thinking of the 'Berserker Hypothesis,' one of the scariest ideas out there. Yeah, John von Neumann came up with the idea of robotic machines designed to replicate exact copies of themselves. Later others developed the concept into von Neumann Probes, spacefaring machines that could scavenge resources wherever they found them and build identical copies of themselves. Each copy would then go out and do the same elsewhere. So what if an advanced alien civilization built such machines that could travel to other stars to nest in suitable planetary systems, find materials to build factories and produce thousands or millions of copies of themselves. They could be thought of as a new kind of life. And what if the alien creators of these probes designed the machines for one main purpose other than reproduction—what if they designed the machines to destroy intelligent life wherever they found it? That way the alien creators could make more room for themselves and protect themselves from possible future threats. For all we know, we could be living in a fool's paradise right now. In another part of our galaxy such machines might be expanding their territory. The killer probes might even have evolved through making imperfect copies of themselves, and some of the mutated probes might have already destroyed their foolish creators."

"Gerry, you're the only person I can talk with about things like this. Other people dismiss such notions and would want to repress them anyway. People gauge risk from tornadoes, earthquakes or wars. Even very small risks like a strike from a civilization-ending asteroid are considered reasonable to think about. But practically no one would bother to think about the risk of extinction from an alien killer probe—it's literally too outlandish. And yet now, here we sit knowing that intelligent extraterrestrials were close enough to notice our radio signals and travel here to stand on our doorstep."

"Yes, I know," Gerry said, "and I'm worried. I would rather think about the Transcension Hypothesis—you know, where we wouldn't hear from ancient civilizations because they evolved so far past their beginnings that they have become almost godlike, would have no interest in us, and have moved on into some other kind of reality that we can't reach."

Jim shrugged. "Yeah, that idea fascinates me too; more even than any of the others. But I guess our extraterrestrial visitors are not like that or else they wouldn't have bothered to come here."

"Well," Gerry said, "let's hope they've evolved closer to the godlike kind of alien rather than to the killer kind."

◈

After Gerry went home, Jim remained troubled. He hated to think about such nightmarish scenarios. But as an evolutionary biologist, he could not escape the logic of the Dark Forest Hypothesis.

Jim remembered the offer of friendship in the message, but he had to wonder if it was a ploy. And he had to wonder by what standards the aliens would judge humanity. It all depends on what they are like. Would they be animals like humans—intelligent but trapped in bodies and brains evolved to survive in a competitive world of struggle? Would they be driven to eat, grow, and reproduce—would they see humans as weaker animals to be exploited in some way? That last thought, the pattern he knew so well as a biologist, could not be dismissed.

But then, what if they had evolved long past their animal origins—what if they were advanced cerebral entities, near godlike creatures that might see humans as unworthy of their consideration. They would notice that humans are the top predators on Earth—the tyrannosaurs of today. The aliens' technology might be so advanced that they have found better ways of sustaining themselves than killing and eating other living things on their world. If so, would they be appalled and repelled by how humans live—would they see humans as primitive and savage?

When he went to bed, sleep eluded him while he listened to Ellen's regular breathing next to him. He looked at her in the near darkness and happened to think how it had been years since they discussed starting a family. Both of them struggled so long with finding good jobs, and they had always been so poor that they hadn't felt ready for that step. Now at last they had steady incomes, but their work demands left them with almost no free time or energy. And anyway, would they want to start a family in this now changed world?

Jim turned and studied her sleeping form in the faint glow from a street light coming through a dislodged curtain. Familiarity had not dimmed his feelings. She was still as beautiful to him as when he first saw her. Sometimes his heart ached with how much he loved her.

Melancholy thoughts persisted keeping him awake. He couldn't dismiss that the aliens might be a threat to all of them. In the natural world, confrontations between the stronger and weaker have a predictable outcome. The aliens might turn out to be the vanguard from an immensely stronger civilization. And there was no comfort to be found in Earth's history of how stronger nations treated weaker nations. Humans had asserted their mastery over Earth, but now they might be facing a force far beyond themselves.

His mind turned again to the idea that the aliens might be so advanced as to be repelled by primitive humans and might judge them harshly. Scenes of predation and violence came unbidden to his mind. An unwelcome memory unfolded, a film he had seen of the dolphin "harvest" in Taiji, Japan—the drive of wild dolphins into a small bay where they could more easily be killed for their meat. For some people not involved and watching from the comfort of a prosperous, secure life, it was a horrifying, violent, bloody spectacle—the hunters' boats floating in sea water bright red with blood and foaming red with the thrashing of dying bodies.

But Jim could not judge the hunters harshly. He well understood that his and Ellen's veganism was a luxury available to them from the abundance of modern agriculture and modern commerce. If they had been born in another time and place—say in an Inuit community three hundred years ago in the far northern arctic—he knew in that case they would be eating seals like everyone else. He knew too that dolphin species themselves worked together to drive fish schools into confined areas to catch and eat them. And pods of orcas, the world's largest dolphins, had been seen herding entire pods of narwhals into confined pockets of coastline to slaughter them all.

Jim could not condemn any of them. They can only be what they are. It's the way of the world. He just wished the world somehow could be different—one that could still function without all of the constant killing. He knew his feelings were irrational. He was a scientist, a specialist in the logic of evolution and knew that nothing would work without the regular intervention of the Grim Reaper. It's a constant battle between life and death. Death always wins in the end. Life can have temporary victories but only when it first pays Death with many victims.

If the aliens were far, far beyond us, Jim wondered if they still could be as understanding of life's struggle for existence as he himself was. *Could they still hold a positive regard for humans while seeing them as primitives—would they recognize humans as a young species but one having potential for good? Would they be able to consider humans as friends?*

He knew that he was able to feel that way about others. He remembered Jube, his dear old dog from the days when he and Ellen were first together. They were living in an old doublewide trailer twenty-five kilometers from the nearest town and on the edge of the Columbia National Wildlife Refuge in central Washington state. Their home was idyllic and magical, placed on a high plateau with dozens of kilometers of scablands stretching out below. Great flocks of migrating birds were attracted there by the many pothole lakes scoured out of volcanic rock by the great Spokane Floods fourteen thousand years ago as the immense ice fields melted at the end of the last ice age.

They were walking there one spring day with Jube when Jim turned over one of the rocks lying all about to see if any interesting insects were beneath. To his and Ellen's delight they saw a grassy nest filled with baby field mice all still with their eyes closed and fur only starting to thicken. The baby mice were squirming at the sudden light and cold. But in a flash before Jim could even think to react, Jube was there gobbling all of them, chewing and swallowing them like candy. Ellen shrieked, and both of them were horrified. But Jube was Jube. He was accustomed to receiving treats from them, and he was a carnivore. How else could he be? He was happy-go-lucky, unfailingly affectionate, never held a grudge—always their beloved friend. Their grief was pure, deep, and long when he died too soon of old age.

Still awake, he remembered one time when he was a student thinking about the whole pageant of life on Earth over the eons. It occurred to him that he himself only existed because there was an unbroken chain of living cells connecting him and some progenitor cell living almost four billion years ago. An absolutely unbroken chain of life! It was "a case of the bleedin obvious" as they might say in Australia, but at the time it seemed to him somehow strikingly profound. And now it felt to him like a desecration to break that continuous chain—that long, long branch of the ancient tree of life on which grows every living thing.

He stared at the dim ceiling listening to the hum of the refrigerator in their small apartment. Then as if a perverse and sadistic director were in charge of his mind, his memory jumped to yet another film he had seen years ago. It was a video taken in an industrial scale chicken hatchery showing skilled workers picking up just hatched chicks, checking each to note its sex and then tossing the female chicks onto one conveyer belt and the male chicks onto a different conveyer belt. Modern chickens have been bred for specialized purposes. Chicken breeds intended for eating grow fast and on minimal amounts of food.

Chicken breeds intended for egg production produce more eggs for the amount of food they consume, but their bodies do not grow quickly enough to be profitable for meat production. And needless to say, males lay no eggs at all so male chicks from egg-layer breeds are not wanted. The video running through Jim's mind showed the conveyer belt with the male chicks taking them to a point just above a large industrial meat grinder where the little male chicks were dumped still living and confused into its roaring throat.

When he first saw the video he felt sick to his stomach and thought—*welcome to Earth little fellas. Sorry you'll never have your chance to sing to the morning sun.* But he knew this was the way of the world. He knew what Earth was like. He wondered—*would this routine, casual killing be a stretch of understanding too far for highly advanced aliens to accept?* He had seen such things too often before in a past job—agitated cattle calling fearfully to each other as they stood queued before the slaughter house door and caught the scent of blood wafting from inside. *Will the aliens look at humans the way humans look at animals? Will they value us in the same way? Will they treat us the same way?* From all he knew of Earth, mightn't that be the case? It was abundantly clear that innocence was no shield. And anyway, how could humans ever claim to be innocent?

Jim wondered whether he was just crazy.

At last he began to feel sleepy. From somewhere a deep half-memory formed in his mind, part of an old epigraph he once had seen in some now long forgotten book.

Dark is the night of our despair when the gods remain silent.

Finally he fell into a dreamless sleep.

Chapter 4

THE VISITORS

Canberra, Tuesday, 23 January, 9:00 p.m.

After running continuously for more than three weeks, the alien radio signal abruptly ceased. The alien broadcast frequencies were being fed into worldwide digital networks so people all around the planet could monitor the message—or whatever might come next—twenty-four hours a day. Minutes of silence passed into an hour while anxiety grew in billions of human minds. Telephone circuits and internet servers around the world were soon overwhelmed. News programs in every language talked of nothing else. Irresponsible "shock jocks" were predicting terrifying possibilities. Religious leaders exhorted their followers to remember their faith and to pray for peace. Government officials urged calm while ordering off-duty military personnel and police to report to their stations.

After two hours of silence, a signal appeared from the sky at midnight Universal Coordinated Time over the International Date Line. People again heard the gentle, reassuring voice.

"Hello people of Earth. We are happy that you have responded to our invitation with so many thoughtful questions. We will begin our answers first to the questions asked most often. We hope that as we progress, you will be further assured of our friendly intentions. We invite you to continue to send us new questions as before. Because we now have an abundant number to answer and expect to receive more, we will begin to send you our answers continuously. We will group related questions together with their answers and repeat that group over the following twenty-four hours before beginning a new group that also will be repeated over the following twenty-four hours."

Again the aliens broadcast in the same eleven most commonly spoken languages with accompanying text scrolling on screen. The gentle, soothing voice pronounced each of the languages perfectly. As with the first message, the natural and error-free broadcast baffled linguists. Although some still thought it must be a colossal hoax, Professor Yang's description of how the aliens could have mastered the languages persuaded most others. And numerous probes from Earth's powerful radar installations capable of tracking distant asteroids had turned up nothing in the direction of the signal. Most important, the now verified fact of the glacially slow movement of the signal across the background star field—movement at just the right speed to match the aliens' claimed orbital position—put to rest the idea of a hoax.

"We first address these related questions—who are we—what do we call ourselves—where is our home.

"We evolved on our home planet originating in a water environment just as life did here. Our science shows us to have developed according to the process you call natural selection, but our planet and Earth each has its own unique history. Our home world has a surface gravity 1.28 times greater than that of Earth's, and that difference influenced our physical structure and made our eventual escape into space more difficult than your experience with Earth's gravity.

"Our home planet orbits a star a little smaller than yours, one nearer the boundary between your classifications of G-type stars and K-type stars. Our planet orbits its star at a distance that maintains surface water in a liquid state, but it is still far enough away so as not to have become tidally locked. Some of our evolutionary history parallels yours, but there are significant differences too. We are carbon-based as you are. The biochemistry of life on our home world is similar to yours in broad strokes, but with many differences in detail. Our metabolism exploits energy-yielding chemical reactions similar in some ways to yours but also with many differences that are unique to us. We can explain more of this later.

"Our home system is located 243 light-years from Earth along the length of this galactic arm in a direction inward toward the galactic center. Your own science has already located and classified a great many planetary systems among stars nearer to you, and our home system would be judged unremarkable.

"As to what we call ourselves, it does not translate into any of your languages, but perhaps the closest term is 'The People.' This choice, though, would not distinguish us from you. We would be pleased, therefore, if you used a term suggested by one of your own on the first day after we announced ourselves. Please call us the 'Visitors.' It is a very appropriate name because we can remain here only for a short time longer and then will need to continue our voyage."

This first answer detonated in the mind of each person who had ever thought about such possibilities, and Jim and Gerry were among those billions of people. Thoughts and questions—thoughts and questions. *Home, 243 light-years away! They're carbon-based just as we would expect as most likely from a liquid water origin. They won't stay long. They even noticed what someone said on a TV program the day after their first announcement!* Listeners hungered for more detail and answers to the many new questions that burst into their minds. But they had to wait while struggling to focus on the gentle voice and what would come next.

The Visitor's voice continued.

"Many of you asked how we know some of your languages so well. Some suspected that we must have been watching you a long while and asked how long.

"We are on an extended voyage of exploration into what is for us a new sector stretching from our home in the direction of your system. While at previous stops along our way, we probed ahead in parts of this region and identified stars with interesting planetary systems including yours. In one important observation, our spectrographic readings told us that your star had a planet that might harbor life, and we scheduled your system as one we would visit. After we had traveled 239 light-years from our home, we observed something very exciting—patterned radio signals from a source ahead of us—signals that were not natural. We confess that there was jubilation among us over this discovery because one of our primary goals is to search for centers of intelligent life. In our explorations so far, intelligent life is rare. Our explorers have found only one other world where it has evolved. But that life has not yet developed a technical, machine-aided civilization. By contrast, worlds with simpler life forms are less rare though not common. We are always happy to discover simpler life forms too, but they are almost always still at a unicellular stage. Although these examples are of great value for study, they do not compare to the excitement of finding intelligent life.

"So you see, these signals were of tremendous importance for us—the possibility not only of a new center of intelligent life, but also the first example of another technical civilization.

At this time we were four light-years from your star system but were not yet certain which planet was the source of the signals which were weak and intermittent. By the time we drew closer, they had grown more frequent and varied. We realized that we had detected some of your earliest radio transmissions that had significant power.

"We still marvel at the extraordinary coincidence of our being so close at the time you first began to use radio technology. We saw that we would have the opportunity to learn more about your development from an early stage. That is how we arrived at your planetary system and confirmed which planet was the source of the signals. We arrived in your system in early 1916, as you number years in your current era, and we have been learning ever more about your civilization since then.

"When we first arrived, we approached your star closer than we are now but remained outside the orbits of its outer planets. We sent observation probes ahead to learn more. Your radio technology soon improved significantly, and we were able to analyze enough transmissions to decode some of your languages. We learned that you had not yet achieved the ability to leave your planet but had developed nonbuoyant machines that could fly through your atmosphere. We also learned that a great conflict was underway on the western edge of the largest continent. Learning of this war was a painful and disturbing experience for us. We have never revealed our existence to the other example of intelligent life we know of. We felt it was imperative that we also protect you from knowledge of us and not interfere at your stage of development at that time. Therefore, we remained hidden and endeavored to learn more about your civilization including why you still engaged in wars.

"For this particular answer it is enough to say that from that time until the present, we have learned a great deal about you. After your invention of television and the continuous broadcast of every kind of program material, we were able to learn far more nuanced detail about your social interactions, your values, your aspirations, and your daily life events. Later, after your invention of computers and your connected internet, our still-hidden probes with their own onboard computers were able to track and interpret your technology as it matured and were able to maintain connections into your digital systems. Since then, we have examined every page of written and photographic material encoded within your internet servers. Indeed, we have studied every scientific paper, scholarly journal article, newspaper article, and every current book as well as the extensive historical and literary collections available within your internet. In addition we have studied every television program, every cinematic film, and every documentary distributed since the beginning of these industries on your world.

"When we first arrived here, we had no expectation that we would be able to amass such a vast amount of information so quickly and easily. In doing so and observing changes you have made during the last 136 of your years and in considering historical changes you have effected during the last 500 years, we now have great hopes for your successful transition to a more harmonious civilization in the future. This last point is a large topic that we can discuss much more on another occasion."

After these words died away, the message began to repeat. It was as if the Visitors knew in advance that these first answers alone would occupy the thoughts of every Earthling for the next twenty-four hours. When the message began to repeat and Earth's listeners realized that nothing new would come for the next twenty-four hours, the human response around the planet built to a frenzied tumult even greater than when the first message arrived. Every commentator, newsreader, and pundit went hoarse describing the reactions of governments and populations across every continent. There were drone videos showing surging crowds in numerous public arenas from St Peter's in Rome to Tiananmen Square in Beijing to Times Square in New York City. Communication systems were again overwhelmed as most of Earth's population sought to talk with their friends and their family.

Before that day, no one knew when the aliens might begin to answer questions, but millions of people had been standing by waiting minute by minute ever since the required time delay for signal travel had passed. Jim and Gerry were among those millions.

After the last answer ended, Jim called to Ellen: "Forget the phone; it might be days before it's working again. Let's just head over to Gerry and Sandra's right now."

"No need," called Ellen who was looking out the front window. "I see them driving up in front." She watched as they walked up the pathway and thought what a handsome couple they were and how happy she was to see them. Ellen opened the front door and said, "You two must be mindreaders; we were just about to head your way."

"We were quicker off the mark," Sandra laughed. "Watch out for Gerry. He's wound up tighter than a garage door spring!"

"Ha!" Gerry snorted. "My darling wife, I don't know what you mean." He then jumped ahead of her through the front door that Ellen was still holding open and began shouting. "Jim ... Jim, can you believe this? I can't believe it—I mean it's unbelievable—it's beyond unbelievable! My God, it's crazy!" All of them knew Gerry was prone to overexcitement, and today looked to be his best performance yet.

Jim was in the kitchen already heating water for tea and hailed Gerry from there. "I'm with you, buddy. Where do we begin? Just think of it! All that spectacular information casually included—it would take humans centuries to collect it—that is if we ever figure out how to travel to other star systems. Really, just think of it! Simpler life forms are not common, but not rare either. And intelligent life forms sound pretty exceptional. And imagine this—we're the first ones they've run into so far who have developed a technical capacity.

And here I've questioned plenty of times whether we can even call ourselves an intelligent species!"

"Ha, you're right about that," Gerry said laughing. "Just look at most of the TV shows and the political stuff that goes on all the time." Gerry was already sitting down at the kitchen table casting his eyes about for a tray of biscuits or cakes.

Jim noticed and said, "Hang on Gerry; bikkies on the way." Jim had already developed a great fondness for the ritual tea and biscuits that accompanied every proper get-together here with friends.

Sandra and Ellen sat down at the table. "Hey!" Sandra said firmly. "Are any of you aware that it's past eleven at night? Thank goodness I don't have an early shift at the hospital tomorrow morning, but still, let's try not to make this an all-nighter again. Nobody wants a doctor in the emergency ward who's falling asleep on her feet."

"I wonder," Ellen joked, "if we could persuade the aliens—I mean the Visitors—to change the time they begin their announcements. We're not natural night owls. The people in America and Europe are much luckier with the timing for them. Maybe we'll turn into night owls or else learn to be patient enough to wait for a nice breakfast meeting the next morning."

"Nah, that'll never happen! We could never wait that long!" burst Gerry. He paused. "But yes, I should be more considerate of my lovely doctor here," he said more quietly as he patted her arm. The look from Sandra made it clear that she put little stock in that.

Gerry, back to loud again, said, "Just imagine all the new things we have to think about just from what we heard tonight. My God, they say their home is 243 light-years away! How could they travel that far? What kind of vessel must they have? Are they descendants of the original explorers who set out who knows how long ago or do they just live an incredible length of time? Any half-way feasible technology we imagine today would take several thousand years or more to get us that far. And get this—they talked about making stops along the way. My God, they're talking like sightseeing tourists! Hey, let's stop here to have a look at this weird planet. And then tomorrow let's pop over to that huge double-star system. For us, that would be crazy! If we could ever marshal enough power to build the velocity needed to travel that kind of distance in any kind of realistic time frame, we wouldn't dream of throwing it away to decelerate for a midvoyage stop. It would take an immense amount of energy to decelerate, and then you would need that same immense amount of energy again to build the velocity back to continue the trip. We have nothing that would allow us to make even a single leap."

"I hadn't thought of that. It's so beyond us," Ellen said. "I wonder if there is any chance they might share some of their technology with us."

"Just think of how they could help us with medicine!" Sandra said, her excitement showing too. "They're carbon-based like we are, and even though their biochemistry would be different in lots of details, it must still follow principles we would understand. And heavens, they have studied every scientific paper published here! Can they mean that? Are they far more intelligent than we are? If so, wouldn't they see things we've missed? It could be wonderful."

Ellen jumped back in. "Yes, that rattled me. They say they've read and studied everything published and filmed since they've been here plus older literature and historical writings too—all that's somewhere in the internet. If they're telling the truth, they know far more about us than any one human knows. We don't know what their individual capabilities might be, but just from the fact they're here, we can guess they know what they're doing."

"You know," Jim said, "I'm wondering if we're all in some kind of denial. We seem to have swallowed wholesale everything they've said. It's all so exciting and amazing. I would love to think that their advanced technology points to their being advanced ethically too. But from Earth's history we know of many militaristic states headed by tyrants that developed advanced technologies to be used for war and conquest. We don't know if the aliens cooperated voluntarily to achieve their technology or did so under the lash of a ruthless dictator. Or they might be pirates who stole the technology from another civilization and are just setting us up for some kind of raid. Or it might be even stranger—they might be like hive insects with some kind of group mind and not able to appreciate beings like us." Jim was rambling now.

Irrepressible Gerry pulled them back to excitement for a few minutes, and then Jim countered with more dark possibilities. Like a seesaw, back and forth ruminations went on for another half hour carrying them past midnight before Sandra herded Gerry, still talking, out the front door to head home. Jim and Ellen were keyed up but grateful to go to bed at last.

❖

Washington, DC, Tuesday, 23 January, 7:00 a.m.

In the US capital, it was barely light on a dreary winter morning when the first set of answers began. A feeling of semicontrolled chaos filled most people's minds. Civil authorities struggled to maintain order after the disruptions during the two-hour period of silence from the aliens. Almost everyone who understood the eleven languages stayed awake to listen to the first round of answers.

By the time the first two answers began to repeat, people began moving again. It was the same in most cities around Earth. People had a range of reactions. Some sat alone and thought, some prayed alone or with family and friends, some joined friends for agitated discussions, and large numbers roamed the streets talking with anyone they encountered. Those who had to report to work did those things with colleagues when they could. The military, the police, first responders, medical emergency personnel, and other vital workers maintained enough discipline to function though with distracted minds. Officials and public service employees in the numerous office buildings of Washington, DC, put on a good public face appearing to carry out their duties but with thoughts as far away as everyone's. Nothing felt as if it would ever be normal again.

President Kaitland knew the coming day would be a marathon of meetings and bedlam. A month ago she had been focused on her upcoming State of the Union Address and the fact that 2052 was the year she would need to campaign for a second term. Then came the message. She and other national leaders were blindsided by something unthinkable. Other leaders in the past had faced existential threats to their cities or countries from invading armies or plagues, but this case seemed far more extreme—a possible threat to all of humanity. No one knew the intentions of the aliens or the truth of their claims.

So it was a day turned upside down as badly as the first day of the message. As the hours passed, billions of eyes looked at clocks anticipating the second round of answers. The crowds that had erupted into the streets after hearing the first set had retreated out of sight. People, pensive and worried, began to huddle together indoors feeling safer under cover. Streets and plazas were abandoned. An eerie and disconcerting quiet descended over public places around the world—an oppressive quiet that persisted like the weight of a held breath beneath eyes fixed on the slowly stepping numbers of a clock.

Chapter 5

SECOND REVELATION

Canberra, Wednesday, 24 January

The next day in Canberra as the clock stalked 11:00 p.m., Jim and Ellen were already settled at Sandra and Gerry's around the TV and flipping between news channels while commenting on every new detail. Earlier that day Australia's prime minister, with the agreement of Parliament, had called all Australian military reservists to active duty, but to do what was not disclosed.

"Well, it might have to do with preserving order if riots break out, but there might be another reason," Gerry said.

"What do you mean?" Sandra asked.

"Well, we know that we are a small nation in population—smaller than Mexico City or any of a dozen Asian cities. We do have strong, well-equipped defense forces, but we're mainly protected by a stable international order and by strong relationships with strong allies. If this new situation preoccupies the US, Britain, and the rest of Europe, there are other powerful nations near us that have bought our raw materials for decades who might decide that now they could do better than continue to pay for them."

"Thanks a lot, hubby," Sandra responded. "I needed something else to worry about."

"Yeah, me too," Ellen added. "I've been trying and trying to counter my worries by wanting to believe the aliens' assurances of goodwill even though I know it is naive to do so. I want to believe that anyone who can develop a technology to travel so far would also have developed a civilization and behavior better than ours."

"I would like to believe that too," Jim said. "Sorry I've been so negative. But we have no way to know whether they're telling the truth. I've asked myself, if they do intend to harm us, why didn't they attack in 1916 before we had missiles and nuclear weapons. But of course the entire story of their arrival that long ago could be fiction—maybe they arrived much more recently. Maybe they're playing us to have us relax our guard because they're not as all powerful as we imagine. And even worse, if they are as powerful as we imagine and are not benign, then what hope do we have?"

"You're not exactly cheering us up, Jim," Gerry observed wryly.

They were a troubled group as the last minutes ticked away toward 11:00 p.m. there and midnight UTC over the International Date Line. They called up the volume on their large screen and heard the first night's answers still repeating. At ten seconds to the hour, the voice stopped and the screen indicated "Stand By." At almost to the second on the hour, new text appeared and the gentle voice began again.

"Hello," said the Visitor's voice. "We greet you again with more answers to your questions. We begin with a group of many questions summarized in this way—how can our vessel have traveled so far—how fast can it travel—are we very old so that we can endure such a long journey or are we the descendants of those who began the journey?

"First, our vessel can accelerate to a nominal cruising speed of a little more than 0.79 of the speed of light. To do this, it does not use mass-reaction engines as your rockets do. Your own science has discovered a number of fields that permeate our universe but has not yet discovered all of them. We have learned how to exploit one of those fields yet unknown to you in a way that can propel our vessel to high velocities. It requires interacting in a particular way with what your science has called 'dark energy' in some contexts or 'vacuum energy' in another. Your interpretations are not developed far enough to see the connection. Indeed, we too have more to learn about this subject, but for now we have learned how to use it for our long range vessels.

"A crucial point is that at the end of a journey, we are able to decelerate to a stop by feeding the relativistic energy gained earlier during acceleration back into the field—repaying a debt to the field so to speak. Therefore we are able to undertake long voyages that can include intermediate stops along the way. There are inevitable inefficiencies arising from mass displacement of interstellar particles by our ship and unbalanced interactions with distant but perceptible gravitational fields. We deal with these inefficiencies by carrying a supplementary energy source not only to power vessel systems but also to supplement our drive engines as needed. This, of course, is a superficial description.

"The need of a supplementary energy source, is the actual limit on the ultimate length of voyages we plan. Between the events of acceleration and deceleration, our vessel coasts just as your rockets do between thrust events from their engines. But during those intermediate periods, supplementary energy is needed. If that were not the case, we could travel distances limited by our personal ability to endure the time of the voyage. We are still working on this problem, but for now we must submit to its limitation. With our current technology, we are able to undertake voyages that in 'external time' can endure more than 4,500 of your years. As you are familiar with relativistic time dilation, you know that the time for us on board the vessel is shorter than for our friends at home. Still, it is for us a lengthy commitment with our velocity being only 79 percent that of light.

"We do need to emphasize that we are bound in space and time just as you are. When we undertake these voyages, we scale them to be consistent with our goals, which do not include seeking new planets to colonize. We do not send vessels out never expecting them to return. Our aim is to learn, and all of our expeditions have departed with a scheduled return date so that new knowledge can be better shared. That is one of the reasons we will be leaving you soon and traveling home. Our journey home will be a happy one as we anticipate bringing news of our special discovery—a new intelligence and our first with a nascent technical civilization. The other reason relates to the length of time we have been away from our home and our friends. We miss them and the joys of home which is no longer the planet of our origin. Our home now is a region of space around our home planetary system and includes three other nearby stars and their planets. We build and live in space habitats that either orbit a planet or a star in that region. We find living in them preferable to living on the surface of a planet.

"We are eager to return home to share our experiences. This voyage has been an extended one and a fruitful one. The time we have spent here with you added to the time we spent at previous stops and the time needed to return home will have kept us away more than 2,450 of your years in the external time of our home. We do have longer life spans than you, but this is still a significant length of time for us. It would be comparable to one of you taking a voyage of a bit more than two of your years."

At this point, Jim, Ellen, Sandra, and Gerry sat in stunned silence. Everything they had heard was staggering. The scale of the Visitors'

SECOND REVELATION

life spans seemed impossibly long. Again their minds raced, and the hunger for more detail was almost unendurable in the face of these dumbfounding revelations.

"The second question is another that was asked many, many times. It is—what is our appearance—what do we look like?

"Having learned much about you from your broadcasts and literature, we understand that this is an important issue for you. You are primarily visual in your perceptions, and your emotions can be triggered by visual appearances. We mention this first to ask that you broaden how you might view us. We are not similar to mammal species on your planet that you might find attractive.

"Like you, our distant ancestors first evolved in a marine environment. But unlike the bilateral symmetry of the dominant vertebrate and arthropod forms on Earth, our dominant life forms followed a radial symmetry as seen in many of Earth's marine creatures. When our ancestors emerged from the sea to be land dwelling, they moved into an atmosphere partly containing oxygen that had already accumulated through the action of our analogs to your plants and photosynthesizing microbes. That oxygen atmosphere was critical because after we had developed sufficient intelligence, it made possible our discovery of the simple energy source of combustion. From that beginning, like your myth of Prometheus, we were able to begin a development similar to yours by using heat from fire for everything from refining metals to powering steam engines.

"Through this long period on a planet with a greater surface gravity than yours, you would have seen us as somewhat squat beings with six legs around the circular base of a cylindrical trunk and with six arms arranged around the top of the trunk. Although our bodies are of a different pattern than yours, they are supported by endoskeletons just as your bodies are. Our brains are located in the upper part of the trunk under the arms and in a protective carapace of bone. We have multiple visual organs distributed over each of our arms. These make possible the blending of a large number of visual fields and gave us the ability to master our surface environment as well as look upward in wonder at all the lights we saw in the sky. Our vision is sensitive to a range of wavelengths similar to yours but including a little more into the infrared—an ability we acquired after emerging from the sea. We have other senses like yours as well, but unlike you we also have a sense that perceives changes in electric and magnetic fields. That sense has been very important for us, but still we mainly rely on visual perception.

"The earlier time we just described we call the time of The Old Ones. Since that time we have changed our appearance and although it remains similar overall, it is different in some details. After our science developed to the point where we could manipulate our genetic information, we began to address some of our problems by altering ourselves with changes to our genes. We undertook a directed evolution over a long period that has brought us to our present state—a state where we have reduced biological bodies that are linked with auxiliary machine bodies of our own design.

"We have designed these machine bodies in many different forms each optimized for different functions. These machines are extensions of our brains in the sense that we can manipulate them directly with our minds. They feel to us as we imagine your bodies might feel to you. You might attempt to imagine how it is for us by thinking, as we know from your literature, how you sometimes feel that controlling an automobile or an airplane can feel as if it is an extension of your own body. For us it is even more direct than that.

SECOND REVELATION

"Finally, to add context that might help you understand how we can be happy having given up the planetary environment in which we evolved, we ask you to consider this: Think of how you perceive yourselves as being somehow behind your eyes. Indeed, your consciousness seems to exist behind your eyes and is connected to many kinds of nerve receptors throughout your body. In a sense, you are a prisoner in your bony skull and informed of the outer world by means of biological machinery that uses biological sensory transducers. You have learned to love the beauties of your planetary environment, and we agree that there are many—ones that your sensory organs perceive well whether the beauty of a brilliant sunset or the feeling of a cool breeze over your skin or the excitement of a deafening thunderstorm.

"Now imagine that your mind is integrated into an artificial body that has new and enhanced senses that your biological body could not have. Imagine flying through space at will seeing different planets with unique environments and exploring them without being limited to the narrow range of electromagnetic wavelengths that is your natural visual spectrum. Imagine 'seeing' with all the wavelengths you know of and more. Imagine sensing natural phenomena in ways you have never experienced. Imagine being able to swap into different kinds of bodies as needed to explore those new environments. We assure you that you would fall in love with many new kinds of beauties as we have."

A few moments of silence followed, and Jim wondered whether this was the end for this session. All four friends already felt an overload of information so unfamiliar and so astonishing that they struggled to organize their thoughts. It was the same for all listeners on Earth. Scientists were delirious with excitement over the breadth of the implications being revealed. Their excitement banished any caution from their minds over the question of the truth of the alien account. Scientists within security and defense departments of national governments, while just as astonished, were not so swept away. They listened critically always looking for a slip or a clue that might indicate fraud. But the few moments of silence were short, and the gentle voice soon resumed.

"Now we will present one last question and answer for this session. A great many questioners asked this one in many ways. We condense it to this—do we believe in a supreme being that created our universe? In your languages that being is known by different names such as God, Allah, Yahweh, Vishnu, Tengri, the Great Spirit, and many others including names from extinct past cultures.

"In a way our situation is similar to yours. Some individuals among us do believe in such a being, and there are those who believe that such a being does not exist. Others of us take a position that you would call agnostic; this is the position of the majority. For those of us with this view, it means that we do not know how we could ever determine with any confidence the answer to the question of a supreme being's existence. This means that the agnostics among us neither believe that such a being exists nor that it does not exist. It is important to emphasize that the agnostic position is not one of mere indecisiveness. If in the absence of verifiable evidence one cannot be confident one way or another over a question of knowledge, then it is more honest to face that uncertainty and learn to live with it until perhaps evidence does become available.

"This is the position of the great majority of our specialists in science. As we have worked to comprehend how the universe operates, we have learned that it functions according to principles that are reliable once understood. It might be that there is a creative intelligent entity behind the orderliness of these natural principles, but we have no particular evidence supporting this possibility. Moreover, we do not know how to observe or interrogate things outside of our universe. As scientists we do ask ourselves what caused the universe. But to speculate what came before a presumed beginning remains imaginary without access to evidence. If we try to answer our questioning minds by saying that before the beginning there was the supreme being who created all that we see, then we find ourselves also asking who or what created the supreme being. At some point we are inevitably blocked without an answer. We choose, therefore, to stop at the point to which evidence takes us. But we always keep our minds open to any possibility from new evidence.

"We have studied your progress in theories of fundamental properties of the universe and have explored similar approaches such as the possibility of higher dimensions of reality. For instance, we have considered quantum entanglement as a hint that two separated locations in our three-dimensional space might be directly connected in a higher dimension. The mystery of questions like this keeps us humble in the face of the overall mystery of the universe. We have learned to live with uncertainty in many areas, and the question of a 'first cause' or 'prime mover' is one of them.

"We continue to search for answers. In the end, those of us following different persuasions on the question of a supreme being live together in harmony and are not troubled by our different views. We recognize that faith and knowledge need not cause conflict between those of goodwill. And we all understand that our curiosity, our willingness to help each other, and our systematic science together have provided us the means to travel from one star to another and return again safely home."

At this point, the voice faded away for a few moments, and then the message began to repeat.

Gerry switched off the receiver and sighed. They sat in silence for a few minutes each attempting to order their thoughts and failing.

Finally Gerry said, "Their life span ... I don't know what to say. Depending on what they meant by 'a bit more than two years' and how long they spent in time-dilation at their higher speeds, their life span is around ninety thousand years or a little more. How much knowledge could each of them have accumulated?"

Jim said, "Whatever happens from now on, our lives will be different than we had planned." Fatigue weighed them down, and they decided to meet again the next morning to talk more. They said their goodbyes, and Jim and Ellen went home.

◈

As the hours wore on after the message began to repeat, the reaction of humanity was different than it had been the day before. Everything was much quieter and restrained. People were shocked by the first set of answers, overcome with so much unexpected, staggering information—especially the revelation that they had been under observation so long and so intensely. The second set of answers contained information richer and in ways more profound, but it also prompted many to feel small and insignificant compared to the aliens. Their astonishing life spans, their accomplishments, their breadth of history—all together they seemed overwhelming. And yet with all this, if they were to be believed, the aliens remained humble in facing the mysteries of the universe.

But scientists scattered across all continents were euphoric and felt a level of excitement beyond their experience. The wonders described by the Visitors set their minds racing to think how such things could be and how they too might find such insights. Those in fields like biochemistry, genetics, and biotechnology were ecstatic over what they heard—a cornucopia of amazing new possibilities to consider. And the small number of xenobiologists that sheltered in various university departments were in a state of rapture while they savored these new revelations. All felt as if great new doors of discovery had opened to them letting in a brilliant burst of light even while they had no clue of how to walk through those doors.

Later that day in Rome, the pope announced that the Catholic Church had no difficulty in accepting the reality of the Visitors. He said, "If God could create us here, He could create other beings in other parts of the universe. Else why had He made the universe so vast? We can learn many things from the Visitors, and they can learn from us."

Other Christian groups took much the same view and were finding ways to accommodate the new situation and expand their circle of sympathetic regard to other intelligent beings. A few fundamentalist groups had greater difficulty, but even among them there was growing acceptance due to the absence of explicit sacred text that ruled out the possibility of intelligent beings elsewhere than Earth. Some even wondered whether angels described in the Bible might also include certain extraterrestrial beings.

Representatives of Islam too were taking an expanded view of these new beings. They were suggesting that Allah might have created the aliens for a different purpose than he intended for humanity. Leaders of other of the world's great religions made announcements as well with nearly all choosing to accept a changed view of who were their god's creatures without making judgments about the purpose or goodness of the Visitors.

General talk shows focused on the aliens' description of their views of a supreme being—their belief or lack of it. Some commentators congratulated the aliens on their sensible approach; others decried them as unbelievers and still others expressed pity for them as lost souls who were still searching for the truth. Some wanted to discuss the details of their views with them, and others wanted to try to convert them.

By contrast, national leaders were not so accepting. US officials outwardly continued to urge calm, but in Washington at the higher levels of the Executive Branch, the new information made decision makers more alarmed. They divided into two groups—those who believed none of the aliens' answers considering them a means to weaken Earth's resistance and those who did believe the alien's narratives and were affected by a feeling of insignificance. Both groups saw the aliens as a threat to their order and pursued every imaginable strategy to develop some kind of effective defense. But they were still entangled in the recent dramatic revelations and were mired in a quicksand of their impotent options.

❖

Canberra, Thursday, 25 January, 7:00 a.m.

The next morning, Ellen and Jim hosted Sandra and Gerry for breakfast. All were energized and ready to talk. As usual, Gerry was hyperexcited by all the new revelations. "I'm just blown away that they are being so open about their history and how they live," he began. "We know nothing about their psychology; maybe they aren't as fearful and suspicious as we are. If we were in their place, we would be cagey about what we revealed to prevent a hostile civilization from knowing too much about us."

"Well yes," Jim said. "But they know so much about us; it might be that they see us as no threat. Maybe we are simple primitives with no foreseeable power to harm them. And there is still the issue of whether they mean us no harm. Our own history gives us plenty of reason to worry. After Columbus discovered the islands of the Caribbean, other Europeans arrived and worked the entire population of indigenous people to death on some of the islands. And that's only one such example of many."

"Or," said Sandra, "they might have a benign intent but be like European missionaries sent to the Americas and Micronesia several hundred years ago. They might aim to change us in ways that we would not choose, and we might be forced to change against our will."

Ellen joined in with: "Oh my, we'll soon be depressed again. Yes, all of these things are true about how humans have behaved against one another in the past, and tragically some of these abuses still persist. But I like to think that we've made genuine progress since then. There's a growing recognition of human rights and fairness, and modern communication and transportation technologies have made remote parts of the world less remote and less different. I have always believed that as a civilization matures and masters its environment so that its people don't need to struggle desperately each day just to exist, then that civilization will also improve the conventions and laws that help a society function more harmoniously. And that harmony comes from a recognition and observance of fairness, something humans innately understand. I want to believe that the Visitors are better than we were. I want to believe that their system of ethical principles is as advanced as their technology seems to be."

Gerry piped in: "I have the same hope, but I worry that they might be so far ahead of us that they wouldn't recognize us as intelligent enough to deserve rights. What if they think of us the way we think of animals that we use in laboratory experiments?"

"But they seemed concerned about us," Sandra said. "They were distressed by our war when they first arrived. They recognized us as an intelligent species and seemed impressed that we had developed a technical civilization. But ... well yes—we only have their word for that."

"Ellen, I remember one time you were telling me about a school of thought in philosophy," Jim said, "that claimed moral progress is an illusion—that so-called progress isn't real because it can revert and be lost at any time under conditions of stress. Are we assuming too much when we imagine that the Visitors have an advanced ethical culture to match their technology?"

"Oh, thanks, Jim. That's a good point," Ellen said. "Yes there are some who believe moral progress is a myth—that any perceived progress is only temporary and can be reversed at any time because of our flawed human nature. They say that human beings with conflicting needs and desires will always undermine any overarching vision of a utopia. And they support their view with the truth of history—that countless people have suffered and died in the past as golden ages have broken down and fallen into decay. Ethical ideas can be written in books, but in life they exist in the minds of living people and are only maintained through continuous practice. Something that is of the mind and depends on the agreement of others can be forgotten.

"By contrast, progress in science and technology might persist longer because it's anchored in the raw reality of the physical world. So I acknowledge that ethical progress can have reversals and does not always go hand-in-hand with scientific progress. But I don't think the argument of moral progress as only an illusion is useful. In a sense that's a form of giving up—a complaint that doesn't explain much. It starts with a misunderstanding of the word *progress*. The concept of progress is necessarily linked to time. What length of time is the right length to consider? Temporary reversals over a short span of time can vanish when looking at a much larger span where overall progress can be seen. Critics might claim that it's not genuine progress unless it lasts forever. But nothing lasts forever! Any significant improvement that persists for a time is real for the people of that time. I don't have much patience with those who think that because the world is not perfect and because keeping progress on course requires constant effort, then there's no hope that things can be better. That effort to keep improvement on course is part of the necessary metabolism of a living culture. I would be fascinated to know how the Visitors have dealt with this problem. How old is their culture and how long have they worked to achieve what they describe? Have they discovered what is truly necessary for an enduring ethical framework—one that is better than all the many flawed utopian visions we have fought over so destructively in the past?"

"I like that, Ellen," Jim said. "I'm with you on all of that. I think there's no argument that there has been ethical progress over the last

SECOND REVELATION

five hundred years. And people have changed their views over just the last hundred years since we began to understand more of our genetics. We knew early in the twentieth century how human gametes form and how the chromosomes from both parents must be able to pair together for their offspring to be fertile. The fact that any healthy male and any healthy female from any part of the world can between them produce a healthy child should have verified our universal kinship. But now after we have sequenced the DNA of many thousands of individuals from all over the world, it's obvious that all humans are members of the same tribe. The chimpanzees living just in the forests of central Africa have much more genetic diversity than all the human beings on Earth have—we're all brothers and sisters so to speak."

"Jim, you're sounding a lot like a genetics professor," Sandra said chuckling.

Jim laughed. "OK—sorry, you're right. I was heading in the same direction as Ellen and should wrap it up. I just mean that as we've learned more about ourselves, we've widened the circle of those we recognize as family—a greater recognition of other people who deserve to be treated the way we ourselves would like to be treated. And that has stimulated a broader ethic—a growing recognition that cruelty against any living thing that can suffer is wrong. Our history has been to treat living things as objects to be used as we wish and discarded as convenient. But lately we're seeing that we should treat other living things with compassion at the least and for some, afford them protection or even rights. Like Ellen, I would like to think that a civilization as advanced as the aliens have described would also have walked this ethical path and would have progressed beyond us in that way too. But still, for now—yes, I don't see how it is anything more than a hope. I don't know how we can verify the truth of their statements. I don't know if time is our enemy or our friend here."

Chapter 6

REACTION

Washington, DC, Wednesday, 24 January

By the time the four friends had finished their breakfast discussion in Canberra, officials in Washington had worked through hours of intense and contentious meetings. Earlier in the morning after the second set of answers had begun to repeat, President Kaitland met with select intelligence, defense, and security analysts to hear their first impressions. All looked tired having worked through most of the previous night considering implications of the first set of answers.

President Kaitland began, "We all hoped for a confirmation that we're dealing with a hoax. But all evidence indicates the alien presence is real. Dr. Schneider, would you please summarize for us the preliminary assessment you have prepared?"

Dr. Jonathan Schneider, a distinguished looking man in his mid-sixties, was a psychologist with much experience as an intelligence analyst in anti-terrorism. "Thank you, Madam President. We're facing something with no precedent. We have no experience with the psychology of an extraterrestrial intelligence. The possibility that their intelligence is superior to our own magnifies our uncertainty of how to proceed. That said, we can start by asking some questions to look for signs of deception. First, what about the aliens' use of language?

Their words seem natural and consistent with current formal newsreader standards but a little stilted at times or even a touch preachy. We have heard their explanation of how they learned our languages. Does their tone reveal something about them or something about the sources they learned from? Or are they trying to maintain what they interpret as a more proper and formal tone? We can't answer those questions yet.

"It's also interesting that all self-reference in their messages is always plural—always 'we' and never 'I'. That would be natural since the message is from a group with a group's presumed agreed intention. But it might also suggest some kind of group mind. That idea is well outside our experience though it is not improper to consider it in this extraordinary situation. But as yet, we have no way into an insight from that angle.

"Other questions relate to the believability of their narrative about themselves. For example, can our own science tell us whether their description of their vessel's capabilities could be true? Our consulting physicists tell us that we do not know enough to critique the description they have given about their vessel's propulsion. Their description sounds fanciful, but how can we rule out something they say we haven't yet discovered? Their mention of a supplementary energy source that enables them to undertake voyages as long as 4,500 years is plausible with a sufficient mass of certain isotopes, but only if the energy needed does not exceed what could be carried with them. We know nothing of how much energy their systems need, and they have kept all of their description at a superficial level. They're not giving anything away. However, they have acknowledged that they are bound in space and time just as we are and have not claimed anything that violates our understanding of physics. That is a point in favor of their credibility.

"Their description of their evolutionary history is also superficial but is plausible. Their self-directed genetic modification is something we can envision, cannot do yet ourselves, and also cannot rule out.

Their long life spans and their life in space habitats are astounding but seem consistent with the scale of a technology that can achieve the other accomplishments they have claimed.

"The end result of these questions is that we do not know enough to catch them out in a deception. One of the main impediments we are facing is the long delay in communicating with them—almost three weeks between our question and their answer. Moreover, they are controlling which questions they answer. As it stands now, we will get nothing from them for another two-and-a-half weeks except for answers to questions we asked the first days after their initial contact. Those answers could be about anything. They must have received millions of questions. If they were close enough for us to have something resembling a conversation—a short enough exchange time—then we might be able to morph a conversation into a disguised interrogation. We might have a chance of tripping them up.

"But inviting them to come in that close is full of risks I do not need to enumerate. They have kept their distance, and we can only speculate why. If they do know what they claim to know, then they know we cannot reach their present position with our technology. They also must know that we possess nuclear weapons that might be able to destroy their vessel if they were closer. Are they keeping their distance because they know we could be dangerous to them at close range? If they do not possess sufficient power to overwhelm Earth's entire defenses, then at close range we might be able to converse and even negotiate with them on a more nearly equal basis. Madam President, other than that possibility, I regret that we do not have a substantive solution to offer at this time."

President Kaitland was not surprised. She had read the first rounds of analyses and thought about the issues from every angle she could imagine. She sympathized with Dr. Schneider's regret. This was a chess game where they didn't know all the rules yet.

"Madam President, may I offer an observation?" She noticed that it was General Thomas Beckworth. She had heard him speak once at a Pentagon conference and knew that he was well respected in the Space Force.

"Yes, of course, General Beckworth."

Brigadier General Beckworth was a strategic analyst and planner in the US Space Force assigned at the Pentagon. He had expertise and contacts in a range of areas that made him in demand for high-level meetings in the Executive Branch. But this was his first meeting at a table with a US president.

"Thank you, Madam President," he began. "Our analysts in Defense have reached the same conclusions that Dr. Schneider explained so well. We know that we have no power to exert against the aliens unless they are close enough to reach with our technology. If we only have spears against their guns, then they must be much closer. Whatever kind of vessel and defenses they may have, it is possible that we could overwhelm their defenses with a well-designed nuclear armed missile force. It is difficult to conceive of a vessel that could withstand multiple nuclear detonations against its hull. I would like to support and add to Dr. Schneider's analysis by suggesting that we begin serious planning on how to develop an Earth defense system. We don't know how long the aliens might remain patient and sit out there answering our questions. Thank you, Madam President."

"Thank you, General Beckworth. I expect that many in the Department of Defense would support your suggestion. Are there other observations or suggestions please?" She looked around the room.

An intelligence expert from the Army said, "It would be reckless to invite the aliens closer. It makes more sense to let this situation develop as slowly as possible. The long message delay is a good thing. It gives us more time to prepare defenses and strategies. With their long life spans, who knows what they mean by leaving us soon. Maybe to them, soon is still years away. Better to keep them at arm's length as long as possible."

Others jumped in next, some in support and some opposed. Soon the exchanges became heated. General Beckworth spoke up again. "I would like to remind all of us of one thing. You recall what they said about returning home. They said they were going home with great pleasure—they were going home happy to be able to report their discovery of us. Does this mean they have not already sent messages to their home region describing us? But why would that be so? If their description of their space habitat civilization is true, the home region would almost certainly have immense orbiting interferometry-style antenna systems to receive messages from remote exploration vessels. If they did begin sending reports back when they first arrived, the messages have been in transit already 136 years. And I would expect them to send regular updates too. But if all that is the case, why would they say that they want to return home in order to share their new important discovery? The fact is, we don't know the truth of their description of their home region. Maybe it's not as advanced as they suggest. I can't help but wonder whether there might be some reason unknown to us, some technical reason why they can't send such interim messages that far.

"And if that happens to be true, then their home civilization won't find out about us until they return closer to home themselves. Here is my question—does it seem important to anyone else here that we would be much better off if their home civilization does *not* find out about us? Yes, maybe they will find out about us later anyway from our old TV signals, but they won't discover as much. According to radio experts in my group, they would see interrupted flashes of carrier band frequencies with some of the actual information lost in the weaker power spread over sideband frequencies. Later on they would see more powerful radar signals from cold war radar installations, but that would tell them little—nothing like what the present vessel could take home with it.

"My point is that they may never have enormous amounts of detailed information about us unless their vessel returns home or reaches a place where it can transmit home. To consider the possible time frame, imagine that the aliens for whatever reason could begin transmitting home after traveling well clear of our solar system. That would be the worst situation for us because their home civilization could receive all that information in 243 of our years. If the alien civilization had unpleasant designs on us, they could reach us from their home in a little over 300 years more. That gives us more than 540 years to prepare ourselves unless the aliens invent some much faster interstellar drive. If we are to survive, we need to think about our future defense needs in a different way. We need to quit thinking only of protecting ourselves from troublesome neighbors on the other side of the planet and instead begin thinking of protecting our entire planet from alien civilizations. We need to think bigger and more long-range. Intelligent aliens are no longer mere speculation. They appear to be real—"

"General Beckworth," President Kaitland interrupted, "are you suggesting that we consider inviting the aliens to come close and then take some kind of action that might stop the alien vessel from leaving our solar system?"

"Madam President," he replied, "I hesitate to make that suggestion, but I can't deny that it comes to mind. Any decision about a preemptive strike is one of such gravity that it should only be approached with the most extreme caution and deliberation. But as we have already concluded, we cannot hope to have any military effect on the alien vessel unless it is much, much closer. It would seem wise to develop action plans as soon as possible in the event they do, indeed, come closer to us, invitation or not. The preparations and capabilities that we might be able to develop in the next months might make the difference between humanity's survival and its extinction."

At this point, Dr. Schneider spoke up. "Madam President, may I offer a few thoughts here?"

"Yes, Dr. Schneider," she replied gratefully. "I would like to hear them."

"Thank you, Madam President. I can see that General Beckworth has thought about our risks carefully, and I agree with some of his points. But I want to emphasize that the scenario he described hangs on at least one critical supposition. It is that because the aliens expressed a desire to return home with news of our discovery, then it follows that there is a reasonable chance they are not able to radio that news home from here or at least from near our solar system. That is a stretch too far. The supposition does not begin to have enough supporting evidence to justify attempting to destroy or disable the alien vessel without warning or without first waiting to find out more. Wouldn't they have sent reports home from their stops along the way here? Wouldn't they have sent reports home immediately after their first discovery of our radio broadcasts? They described the excitement it caused them. They were only four light-years away from us at that point. If they don't return home, wouldn't their home civilization follow their trail of reports toward us looking for them? They would see our present signals and head directly here. If we took an aggressive action and harmed them, what would their home world's feelings toward us be if they found out what we had done?

"And another thing we have not allowed ourselves to consider yet is what if the aliens are telling us the truth about their intentions? What if they *are* goodwill ambassadors who wish to establish contact and begin a friendly and cooperative relationship with us?" Now Dr. Schneider was raising his voice. "What if they are benign beings who could help us enormously? Should we be thinking of destroying them before even attempting to learn more about them?"

"And what," interjected General Beckworth almost shouting, "if they are malignant beings intending to destroy us without a second thought as soon as they get the chance? What if they consider us a future threat that they must preemptively destroy? They may fear that we could grow much more powerful thousands of years from now. They may want to eliminate us from our planet before we become a threat! And how can we learn their real intention without risking everything to wait and see what they do?"

"Gentlemen, gentlemen!" President Kaitland said firmly. "We're looking at things from all angles here—not making decisions yet."

Everyone was exhausted from the previous night, and tempers were strained. "For the time being I think we have done what we can do in this meeting. I suggest we make good use of the next hours with further analysis. Others are working on strategic planning as we speak. Each section here will have a summary report to me by 1800 this evening. I expect all of us will be working late into the night again. Try to get a few hours of sleep before the beginning of the next round of answers at 0700 tomorrow morning. Thank you."

"Thank you, Madam President," echoed from many voices as they left the room.

President Kaitland remained for a few minutes thinking about the discussions. General Beckworth and Dr. Schneider had raised critically important considerations. She knew Jonathan Schneider as a wise grandfatherly sort of man, a careful and rational thinker who had given her useful advice in the past. She did not know General Beckworth but had heard that he was a highly regarded strategic analyst with a reputation of working well within the chain of command. As she stood to go to another meeting in the Oval Office, she knew she might need to face some kind of terrible decision in coming weeks.

General Beckworth walked away from the meeting, consumed with worry and frustration. He visualized perhaps too well the unprecedented and shocking threats facing them. He was accustomed to dealing with complex strategic scenarios inherited from the cold war days—constant monitoring of early warning systems, decision tree analysis for immediate retaliatory launch of missiles, and even scenarios aimed at justifying preemptive strikes. The great divide between East and West was still in place, and an uncomfortable peace still remained dominated by suspicion and mistrust. Although the intensity of the Cold War had taken a step back from what it was seventy years earlier, the concept of mutually assured destruction, MAD, still persisted in defense strategic planning. Any significant mistake affecting that balance point might lead to cataclysmic destruction of one side, the other, or both.

Both sides still maintained massive nuclear arsenals, and both had improved delivery systems aimed at leaving only rubble and death. General Beckworth's career in the Space Force did not include any direct command function in the US strategic nuclear strike forces. But he rubbed shoulders with those who did, and East-West tensions had spilled over into his past command positions. His training and experience had molded his psyche to look for threats from any direction. Now he found himself facing a possible threat that exceeded—by orders of magnitude—any others he had ever considered. And worse, his best efforts might never be enough to help. He was a good man, but a good man under much more strain than he realized.

As he walked down the hallway, his argument with Dr. Schneider reverberated in his mind. He felt a mixture of both anger and fear. For General Thomas Beckworth, this particular morning marked an unseen danger. His mind was treading too close to an edge—an edge where a small misstep could take him over a hidden inner threshold and into a dark mental labyrinth where reason loses its way.

Chapter 7

WAITING

Canberra, Thursday, 25 January, 7:00 p.m.

Three of the four friends had gathered again for dinner and to wait for the new set of answers expected at 11:00 p.m. Sandra was still finishing a shift at the hospital but would be home by 8:30. The evening news outlined new information, and talking heads kept busy with interpretations and comment. Gerry switched the TV to a science-related program that focused on the alien capability of interstellar travel, their long life spans, and their claim to have part biological and part machine bodies. Responsible scientists held sway for a time, but then the program switched to a collection of UFO enthusiasts saying either "I told you so" or claiming that the aliens had already visited them. Gerry switched it off with a sour comment just as Sandra arrived home from work.

Gerry, Jim, and Ellen took turns filling her in with the latest as she ate dinner. As the crucial hour approached, there was almost a feeling of calm normality. All four anticipated something new and important, but they could not imagine being more astonished by what might come that night than by what they had already heard.

Once again at 11:00 p.m. in Canberra and midnight Universal Coordinated Time over the International Date Line, the previous message ceased and the gentle voice announced the next offering.

"Hello good people of Earth. We greet you for the third time with a new set of questions and our answers. We are looking forward to hearing your responses about the previous answers we have sent, but we won't receive your responses for some time yet. Perhaps in the future we can come closer so that we may converse more easily. The first question we will address now can be summarized as—do we have wars in our homeland region, and does our vessel carry weapons of war?

"We did expect this question and are saddened by the fact that you needed to ask it. But we know your history and current situation and, therefore, understand your great concern. War has been absent from our civilization for so long that if we were in your place, we would not have thought to ask about it unless perhaps much later in the context of a historical discussion. Our present civilization is old by your standards. In much earlier times we experienced the rise and fall of many dominant civilizations with their inevitable wars of conquest. Over time, though, we began to recognize who we were—we were 'The People.' We came to understand that we were all related to each other and all had the same important needs. With progress in science and technology, our world became more connected just as yours is doing now. We gradually learned to avoid violent conflict and large scale wars became things of the past.

"By then our science brought us to the point of being able to modify our own genetic code. We recognized the risks of the technology but also the potential gains. We began with small careful changes. As these continued over long periods, they added together to result in major change just as with natural selection. After we were able to leave our planet's surface and establish orbiting research stations, we learned how to protect our own biological systems from the negative effects of zero gravity and increased radiation exposure. After this we began to exploit our nearby space environment much more extensively. Unlike Earth with its Moon, we did not have the great advantage of a nearby base with abundant resources outside the gravity well of our home planet. Our progress was slow, but in time we reached other orbiting bodies like your asteroids that we could use to build larger artificial environments.

"We measure the beginning of our current civilization from about this time—a beginning that we link to the establishment of a peaceful and stable system of government. Since then we have never had a war. War to us is something we would not contemplate except in the most dire of circumstances requiring us to defend ourselves against annihilation. Our civilization has benefited greatly from its harmony and absence of violence. Our current civilization has endured from that beginning for 161,723 of your years. Our experience of war is far in the past.

"To continue with your question—do we carry weapons of war on our vessel—the answer is no. We have no need of them. However we do carry with us tools that allow us to manipulate external objects to some extent. We need these, for example, while traveling to deflect aside small objects and interstellar particles in our path or sometimes to bring objects closer to the vessel for examination."

On hearing this, there was consternation among listening defense analysts in Washington. How could they have a stable civilization so old that its modern refined form began perhaps even before early humans were beginning to develop the capability of language? Could these tools they mentioned be used as weapons? Are they implying that their vessel is not armored against attack? Are they implying that they don't have the ability to defend themselves?

"The next question we will answer is this—what form of government do we have?

"This question is not as easy to answer. Even though our form of government is linked to a concept you are familiar with, we also need to explain developments we have made to make our system functional. In a broad sense, we would call our form of government a democracy. But it is not a form of representative democracy as prevails on Earth. It is a direct form of democracy made possible by more robust and comprehensive communication systems than exist in your technology. It also required a gradual development of laws and systems that ensured the rights and privacy of individuals connected to such a comprehensive communication network. Over time we also chose to tailor ourselves at a genetic level to higher cooperation and lower self-promotion and aggression. And our longer life spans provide more time for each individual to gain experience, balance and wisdom, a critical requirement for a successful democracy.

"Once these elements were in place, we had the ability to measure each citizen's preferences in real time. This made it possible for governance to be consistent with our population's true consensus. It is important to mention that by then our science and technology had provided for everyone's basic needs, and it is also important that our society is no longer one of hierarchical privilege. People take pleasure in each other's successes, and there is no reason to feel envy. As a result, the consensus we refer to is extremely high by your standards. At times when there is some level of dissention, we have negotiation and arbitration methods that so far have been successful.

This is a brief description of a long and complex history. We hope that it answers the immediate question, and we will be able to elaborate more at another time."

There was a significant pause before the voice resumed.

"Our last question to answer for this session is this—why have we chosen this time to announce our existence to you?

"We mentioned earlier that we felt it important to protect you from knowledge of us. That was our assessment when we first arrived here 136 years ago. Since then we have learned much about you. We were horrified at the level of violence we observed, but we have also seen from your history that violence was much worse in the past. True, your newer technology has given you greater power to destroy, but other technologies have given you a greater ability to communicate and to understand each other. That is the beginning of a better way to live together.

"We have also been impressed with your overall rate of scientific progress. You are beginning an expansion into habitats off your planet, and your advances in medical science are improving your life spans. Your advances in biological science have enabled you to consider directed genetic modification to enhance your lives even more. But the possibility of war is always with you, and international tensions remain high. Due to your changing climate, disruption to the lives of large groups of people will add still more tension. We have asked ourselves whether there is any way we could help you without unintentionally making the situation worse. We have studied this question for years now.

"A number of things we have seen about humans give us hope. Despite your capacity for great cruelty and violence, you also possess a capacity for great empathy and compassion. For example, we have observed how your treatment of other animals has begun to change during the last century—a change away from unnoticed casual cruelty to an awareness of suffering in other species. We have seen how groups of you gather on a beach to push great whales back into the sea to save their lives, and we observed how you celebrate when you succeed and grieve when you fail. This is behavior we have not observed to such a degree in the other intelligent life form we have found. These and other similar examples are what give us hope for what you can yet become. You are a young species, and much still lies ahead of you.

"Please forgive us if this frank assessment strikes you as condescending. We dare express it only as an outsider who has already passed through such times as you are now experiencing. One important difference between us is that no older civilization appeared to us to offer the benefit of their experience. We debated long among ourselves whether we should offer our experience to you. We knew that we must leave soon and could leave you with only a little knowledge of us. We also knew that any significant greater involvement with us would interfere with your own authentic progress. At last we decided that giving you an awareness that you are not alone in the universe and an awareness of what might be possible for you to achieve through cooperation might help you lift your eyes away from your current conflicts toward better possibilities and visions of a happier future. This is the short answer to your question. There is more to say, but that can come at another time."

After the voice faded away, the message began to repeat.

Washington, DC, Thursday, 25 January, 7:45 a.m.

In Washington, DC, the questions and answers were completed by 0745. President Kaitland sat in the Oval Office where she had listened to the broadcast with her chief of staff, her deputy chief of staff and several others. She had arranged a meeting for 0900 with the same group from the previous morning. She told them to arrive with a preliminary analysis of what they heard. The previous two broadcasts had left everyone stunned by the significance of the revelations. But the last answer in this most recent set felt different, more personal. The aliens expressed a concern for humans as if they cared about them. She asked her Chief of Staff, Albert Thompson, what he thought, and he agreed with her. He said he found himself wanting to believe them. She was anxious to hear the reactions of others.

President Kaitland watched as the group entered the meeting room to be sure both Jonathan Schneider and Thomas Beckworth were there. She convened the meeting and said, "Again we face a morning with new information to consider, this time with greater detail of the aliens' society. I would like to hear your opinions please. Let's begin with General Beckworth."

Tom Beckworth was startled that she called on him first, but he gathered himself and said, "Thank you, Madam President. I was surprised at their choice of questions to answer today. They could have avoided these in favor of ones less revealing about their social organization. Because my role requires me never to forget the possibility of deception from a potential enemy, I must always consider the possibility of sinister motives. If their intent is deception, then their expression of concern for our welfare would be logical as manipulation to encourage our acceptance of them as friends. They would know that the way they described their government and how their society is organized would be appealing to us. Their description of a true democracy that respects individual freedoms might be a strategy to encourage us to relax and look at them as ones we could trust. But without being able to question them directly, I do not see how we can separate truth from deception.

"There is another point that has occurred to me. The set of answers they choose to give us each day is limited. Although their answers are clear and even generous with gratuitous extra information, there have been only two or three questions answered with each session, and each set takes less than an hour to transmit before repeating for twenty-three hours more. They must have received hundreds of thousands of questions in the first two days after their initial announcement. True, there must have been tremendous repetition of the same questions, but there must have been thousands of unique questions too. I've wondered why they have parceled out answers at such a slow pace. Are they intending to give us more time to absorb particular kinds of information about them and not be distracted by many different issues? I doubt it. The kind of information we have received so far could be designed to mold our view of them as vastly superior yet friendly and helpful. Again, this could be a clever part of their deception, and I would like to be able to ask them about this more directly."

President Kaitland noticed General Beckworth's hint about inviting them closer. She asked others around the table for their views and heard only small variations of what Beckworth had offered. Then she turned and said, "Dr. Schneider, what are your impressions?"

"Thank you, Madam President," he replied. "I would like to begin by saying that I must agree with General Beckworth's analysis and particularly agree with him on his last point about the rate of their answers. I have wondered about the same thing. I think it is safe to say though that they would want us to have a favorable view of them whether they are telling the truth or lying, but we still have no way to judge their true intent.

"I feel the same frustration at our not being able to question them frequently and receive rapid responses. It's the only avenue we have to form a judgment about them, and the only way we can do that is for them to be closer. I understand that we must not drop our guard and must continue a defensive posture. But at some point, we may want to give serious thought to inviting them closer. They themselves suggested that option this morning. It is difficult to remain so passive. We find ourselves waiting for them to give us any hint of what may yet come while we have no power other than to ask more questions and wait three weeks to see if they might provide a response." Jonathan Schneider was frustrated at how weak his advice sounded. Like Beckworth, it was difficult for him to stand by waiting. "Thank you, Madam President," he said.

President Kaitland listened carefully and concluded that everyone had said all they could for now. "Thank you Dr. Schneider," she said. "And thank you all for your contributions this morning. Please again provide me departmental summaries of your further analysis by 1800 today. This meeting is adjourned."

She left the meeting room still thinking about what Dr. Schneider had said. *How true it is that merely waiting seems unacceptably difficult when we don't know whether we are waiting for something good to happen or whether we are a mouse under the gaze of a cat.*

General Beckworth left the room gratified that Dr. Schneider had supported his views today. Perhaps he could enlist Schneider's help in persuading the president to invite the aliens closer.

In the meantime, he and people all around the world continued to stand by, distracted in all they did and waiting ... waiting.

Canberra, Friday, 26 January, 8:00 a.m.

The four friends in Canberra waited too while gathered the next morning to discuss the third set of answers.

Ellen said, "Do you remember what Jim and I were saying yesterday—that a society so advanced in science could also be expected to be advanced in its social organization? Everything we heard last night was straight along that line. I admit that it sounds too good to be true. If it is, my goodness, we're primitive and uncivilized compared to them. Oh wait, I guess there was never any question of that."

"I agree," Sandra said. "It's wonderful to imagine what life would be like if wars were a distant memory. For my work in the emergency room, I've watched training films showing wounded young men pulled out of battle zones. I doubt any of you would get through the films even part way without tears. My training helped me switch into 'professional' mode and watch in an effort to learn, and then it's still very difficult."

Gerry put his hand on hers and said, "Sandra, I'm lucky you were the one who patched up my leg that day years ago." She smiled warmly at him and gripped his hand.

"Yeah, all of this is good," Jim said. "But we still don't have a way to know how much of it is true. I hope it is, but I can't help but worry. We like to think that we know enough to make intelligent guesses about new experiences and forecast what might come next. But this is so different—so new. I've been wondering whether any of the patterns I know about social organization on Earth would apply to them. Or could they just be something crazy-different from us? Ellen, when you said that they might be highly advanced in social organization, I couldn't help but think about bees and other hive insects. I know what you meant by advanced—peaceful ... harmonious. The problem is that the hive organization seems harmonious but at the expense of the individual. We should be pretty careful about taking any of our assumptions too seriously."

There were nods of agreement and sighs of resignation. Gerry said, "This tension is hard to take. I'm ready for some kind of break—something new that might give us some hope, some relief. I've been thinking about the things they've said related to physics to see if I can find any holes in their story. So far nothing. It's a kind of standoff we've never faced before. All we seem able to do for now is to wait."

"I know," Jim said. "Tonight should bring another set of answers. I wonder how long they will keep it up. They must have received hundreds of thousands or probably millions of questions. I'm guessing that plenty would be about things they don't want to tell us." Conversation drifted off, and they agreed to meet again the next morning.

After Gerry and Sandra left, Jim and Ellen talked about what they needed to take care of before their scheduled return to Arizona. Although the last weeks had been filled with tension, in hindsight time had flown by. They had dropped their original plans for touring different parts of Australia. Nothing else seemed as important as following the news each day with their friends. They had arrived in Canberra two days after Christmas, and their flights home were coming up on 31 January. They found it hard to believe that their time here was almost past.

Washington, DC, Thursday, 25 January, 5:00 p.m.

At about that same time it was early evening in Washington, DC, and almost dark—a bleak, cold winter night lay ahead that offered no cheer or respite from the pressure of the day's events. President Kaitland was in her private study reviewing reports she had received from various other meetings. She hoped to find even one significant new insight but was disappointed—they were just recaps. No one knew what to do next other than wait. Staff were doing their best to maintain a professional demeanor, but it was clear that everyone was under strain and behind on sleep.

Rachel Kaitland, though very effective when leading groups, tended toward introversion. Her mind was always active and needed no stimulation from others to work furiously on a problem. And now she was frustrated. In the past she could analyze and sort competing factors quickly to push her way toward some kind of workable solution, and she was often the first one to get there. This situation was different. So many unknowns confounded the possibilities that it was impossible to rank options. She had even put some top analysts to work trying Bayesian analysis but had heard nothing back yet. That did not surprise her. There was not enough information for a reasonable starting point and any estimated probabilities would be fanciful.

President Kaitland had the wisdom to listen. She always thought her way through issues herself, but she would consider other people's ideas fairly. She had been impressed with comments from Dr. Schneider. He seemed more thoughtful in ways that were not typical of most defense and intelligence analysts. It occurred to her that her defense group could have a systematic bias that might exclude valuable insights. It might help to hear other ideas from someone who did not assume that every unknown figure approaching in a fog was an assassin—someone who, perhaps, could see broader possibilities.

As she thought about who might help her, she remembered one person who impressed her two years earlier during a series of meetings with the Presidential Commission on Ethics in Government—a title too broad to be meaningful, but the current members of the commission had experience in a range of areas useful in policy formulation. The meetings focused on ethical issues of past treaties imposed on Native American tribes in the United States. She turned to her computer and searched for the commission members. Soon she saw the face of the woman she remembered—Ellen Hazelstein, a professor of philosophy at the University of Arizona. Yes, now she remembered. She signaled her aide and asked him to bring her any of Hazelstein's books and papers that he could assemble quickly. That small interruption to her thoughts allowed her to remember that she hadn't eaten all day. Maybe she could fit in a sandwich while she studied more reports.

Chapter 8

MORE TO PONDER

Washington, DC, Friday, 26 January, 7:00 a.m.

On schedule, the third set of answers ended, and a new set began. By now the number of people listening to the initial moments of a new set had declined. Everyone knew they would have a chance to listen at some point during its broadcast and that the media would be discussing it nonstop in any case. For this set the Visitors chose to answer questions about their mode of reproduction. Apparently a large number of questioners were fascinated with the aliens' sex lives and wanted to know whether they had two sexes or more, whether they enjoyed sex, whether they fell in love and married, how many children they had and so on. The Visitors would have had many more technical questions from biologists, but they chose to focus on these general interest ones.

President Kaitland convened her morning meeting with the defense and intelligence analysts to discuss the most recent broadcast. The subject matter of the answers was fascinating to biologists, but these defense-oriented analysts struggled to find details relevant to their concerns. In the whip around the table, most had nothing new and wanted to return to things they had said the previous day. One person ventured an inappropriate joke about the aliens' sex lives, which was met with silence and an impatient look from the president. She had a good sense of humor, but this was not the right moment.

She came to General Beckworth. "Thank you, Madam President," he said. "I'm not sure what I was expecting this morning, but this biology lesson did surprise me. It could be a delaying tactic to divert our attention from more critical issues. It's a fascinating story light in detail that seized the attention of the media and popular imagination. It leaves people with nonthreatening information that would reduce anxiety. Again it could be a strategy to cause us to relax and lower our guard."

Spoken like a true defense analyst and soldier, President Kaitland thought.

Beckworth continued, "I know for a fact that the Department of Defense sent a large number of questions that in a subtle way dealt with strategic issues of concern—issues like the size of their vessel, how many personnel are on board, and so on. We never expected them to answer direct questions on these matters, but we did hope to see some peripheral references to points that would allow us to make inferences. Instead they've given us generalities, some of them comforting and others that accentuate their superiority. The single half-clear answer we received is that while they denied possessing weapons of war, they did admit to having 'tools' that can manipulate objects outside of their vessel. My experts assume those so-called tools could be much more. It's like we're a mouse being played with by a cat. We still have more than two weeks from now before we could receive any responses the aliens might choose to give us to the additional questions we asked after we saw their first answers."

President Kaitland thanked him when he finished. Except for the attempted joke about the aliens' sex life, she was careful not to reveal any reaction when her advisors spoke or in this case vented. She did not want them to feel inhibited by any negative facial expression or vocal response from her—she wanted to hear what they thought without self-editing to please her. As she listened, she noticed that he used the cat-and-mouse analogy that had occurred to her. She saw the strain he was feeling in the information void they all felt. *Is he overreacting?* This was only the fourth round of answers, and she made a mental note to see what her chief of staff thought about him. She moved on around the table.

Dr. Schneider was the only one to offer any substance related to the most recent broadcast. "Well," he began, "we did find out something new; they have genes or the equivalent, and they are not divided into two sexes with different reproductive organs. They're hermaphrodites, but none are able to reproduce alone. Two of them still need to get together and exchange and reshuffle their genes somehow. They didn't offer details, but since they are alike physically with the same reproductive roles, their society presumably would not have the developed rigid gender roles like ours has. That might be one source of tension and conflict they are free of. And they said that they formed lifelong partnerships and used our words 'loving and supportive relationships.' Both of these features lead me to think that their psychology might not be totally alien to ours—that we might have enough in common to allow us to relate our emotions to each other. That thought is consistent with the story they have given us about the harmony in their society. But the biology they described was what they were long ago before they began to change their genetics and bodies so extensively. Now they say they rarely reproduce because their life spans are so long, and their reproduction is no longer strictly tied to their bodies but is done in machine incubators. So it is risky to draw any conclusions about their present psychological nature. We don't have any biologists in this group. I wonder if some of them might have better insights."

President Kaitland was surprised at Dr. Schneider's last comment, because she had already scheduled just such a meeting with a group of biologists for that afternoon. This group of six with appropriate background checks and security clearances included recognized experts in behavioral biology, genetics, and evolutionary biology. They had been selected and organized by executive staff two weeks ago in anticipation of a day like this, but it was an extraordinary coincidence that new information on the aliens' biology had arrived just that morning. Today's meeting had been intended as a first attempt to draw ideas and possibilities together from experts. She allowed herself a small self-congratulation at her foresight.

At 1500, she convened the biologists' meeting with a half-dozen White House staffers also in attendance. After introductions, she thanked the group for coming to Washington and began. "I'm confident that all of you have heard and been fascinated by the information the aliens have given us in the last few days. This morning we were given more detailed information about their biology that I hope might prompt some ideas that could lead to useful insights. I'm sure you appreciate this situation is without precedent, and we are looking for unprecedented approaches. I'm aware that you met as a group yesterday for a first discussion, and I'm looking forward to hearing what thoughts you might have by now. Who would like to begin?"

Dr. Ladislav Kovacs, a tall, white-haired man responded. "Thank you, Madam President. Since yesterday we have all been working together. I was asked to present our thoughts to you, and we are available for questions at any time."

"Thank you, Dr. Kovacs," the president responded. "Please continue."

Dr. Kovacs stood and went to the front of the small meeting room where there was a large display screen. "We have assembled a few visual aids to explain some of our ideas." He brought up an artist's illustration of the ancestral alien form that the Visitors had described two days ago and that had appeared everywhere in the media. The image had no detail other than what little the aliens had described. "The aliens gave us this description of their older form but said that they have changed since then; we don't know how much. They did say that they had already evolved into this older form by the time, in their words, they became sapient beings. This form has no precise analog on Earth. The closest are marine animals like sea anemones, starfish, and jelly fish—none have endoskeletons and all are sessile or slow-moving. Exceptions to this pattern though are found in the Cephalopoda in the phylum Mollusca which has—"

"Excuse me, Dr. Kovacs," President Kaitland interrupted. "Perhaps you could jump ahead to your conclusions so we can discuss them in the limited time I have here." She was not uninterested but didn't have time for a full-blown professorial buildup.

"Oh ... oh, yes, of course, Madam President," Professor Kovacs said a little sheepishly. He abandoned his display screen and turned to the group. "We are guessing this animal had no front or back. They mentioned their visual apparatus as distributed over their six arms, and they were able to blend together many different visual fields. That suggests they could see in all directions simultaneously. We think it could have changed direction while in motion without rotating its body—an advantage whether its earlier forms were predators or prey. Whatever the environment was that molded it toward intelligence, we imagine that because of its visual capabilities, it would have a superior perception of spatial orientation. This feature may have contributed toward their development of intelligence, and we believe they would have retained that spatial talent as they molded themselves into their modern form. From this ability, they might have developed a superior form of mathematics to ours.

"They gave us no detail about how their arms are terminated such as by flexible tips like tentacles or by analogs to our fingers. However, having six arms would be very advantageous for a tool-constructing species—very useful for manipulating objects and perhaps even contributing to the development of more complex brains to control the manipulation. It is reasonable to speculate that their brains have six lobes suggesting stronger multitasking capabilities than ours. Their centrally located brains protected by bone would be well-placed to coordinate signals between the arms and legs—again a possible advantage. We don't know though whether any significant portion of their earlier bodies remains in their present form. We surmise that, in their genetic manipulations, they would have retained any valuable brain and body features that had already evolved.

"The aliens' main emphasis this morning was on their reproduction. From their description we know they have something similar to our genes, but they provided no detail about the nature of their genes, their organization, or even about the molecular structure in which the genetic information is encoded. We don't know whether they use DNA, RNA, or some other similar molecule that has the critical encoding and replicative properties. The fact that all have both male and female reproductive organs and that each can produce an offspring from a mating event is fascinating. We know nothing about their culture in earlier times but guess that in principle they would not have had reproductive role differentiation. It is possible such a situation might lead to a culture with fewer sources of conflict—as in men fighting over women here—and possibly a culture with greater coherence. But this point must remain speculative for now. If we take their descriptions of themselves as true though, they are a harmonious society.

"We noted that they described their reproductive arrangements as based on pair bonds—a close relationship between two individuals that they described as loving and supportive. We don't know whether the offspring are reared as a nuclear family within that pair bond or whether offspring are grouped and brought up together in some kind of communal child-rearing center. We also don't know how long the rearing process requires. But we do think that the description of the pair bond as loving and supportive provides evidence of an emotional attachment similar to ours and, therefore, as something each of us could understand about the other. It also suggests that they might feel other emotions we do such as empathy, compassion, and so on. They have referred often to those emotions. Our science has speculated that these human emotions were refined from those needed for successful child rearing over the long period humans need. It might be the same for them.

"Madam President, since our meeting yesterday, we have considered some rather outlandish ideas about the aliens' psychology and potential for traits such as aggression, retribution, domination, and so forth. But it can only be conjecture without evidence to support it. I'm sorry to say that with the limited information we have now, we can only offer the few points I have just described. We wish there were more and perhaps over the next few days new facts will arrive to give us more to work with. But for now, I regret this is all we have. Thank you, Madam President." He sat down.

This group of experts said much the same thing as Dr. Schneider did just after he first heard the broadcast. President Kaitland was now even more impressed with Jonathan Schneider's insights. Another twenty minutes of discussion wrapped it up, and she adjourned the meeting with thanks. She asked them to provide her a written report with any additional ideas by 1700 tomorrow. She also asked them if they would remain available after they returned home to consider new information and send her reports of developments. They all were well-known and prominent in their fields, but it was unusual for them to be asked to assist the President of the United States. They readily agreed.

Later in the afternoon while thinking over the meetings of the day, she remembered the papers and book of Ellen Hazelstein that her aide had brought her that morning. She reached for them and put them in her satchel for bedtime reading that night.

Canberra, Saturday, 27 January, 8:00 a.m.

By the end of President Kaitland's meeting with the biologists in Washington, the four friends had settled down for breakfast this time in Gerry and Sandra's kitchen to discuss the fourth set of answers. Sandra was still working the noon to 8:00 p.m. shift at the hospital, and had been home in time to listen to the broadcast with Gerry the previous night. This time she managed to get him to quit talking and let her go to sleep by midnight so she was in reasonable shape this morning.

She opened with, "Let's have a relaxing breakfast today. We only have a few more chances like this before you two are on a plane back to the US."

Ellen groaned. "Yes, we don't feel like going home yet. We're happy at home, but we have enjoyed our time here so much in spite of all the worry. And we never managed to explore the other places in Australia we wanted to see. There's no doubt about it—we will have to come back soon. And it goes without saying that you are welcome to come visit us anytime you want! You two have made this time so special for us."

"I'll second that!" Jim added. "I predict we'll be best friends the rest of our days even if we do live on opposite sides of the planet."

Gerry and Sandra beamed.

Sandra said, "This sounds like group-hug time to me, but hey—let's just dive into the fruit salad. Anyone need more coffee?"

"What time is your flight out of here? It's on the thirty-first, right?" Gerry asked.

Jim said, "Yes. We fly out of Canberra at 9:00 a.m. to have plenty of time to get though all the fuss at Sydney International and board our 2:00 p.m. flight to LA. It wears me out thinking about it. I'm always nervous when I need to make an international connection and can't relax until I'm settled down in my seat on the big plane. It's a long hop across the Pacific so I'll have plenty of time to relax once I'm on board."

Ellen laughed. "Yes, he's nothing but fidget, fidget, fidget until the plane starts its takeoff roll, and then he's like a boy on a pony for five minutes until we're well in the air. Then he's back to fidgeting again and chat, chat, chat when I want to take a nap."

"Oh you poor thing," Jim laughed. "What you have to put up with!"

"Well," Gerry said seriously, "we'll miss these times together. It's been just as special for us."

They all nodded and smiled sadly.

Gerry asked, "What did you two think of last night's broadcast?"

"It wasn't as startling as things we heard before," Ellen said, "but it was fascinating. I wonder if the hermaphrodite mode of reproduction is more common on their planet as opposed to what we have here."

"Yeah, that caught my interest too," Jim said. "There are reasons why an asexual mode could have an advantage over the sexual model, but there are also reasons to justify how sex became dominant on Earth through its better maintenance of DNA integrity, its ability to promote variation in offspring, and the ability to better sort advantageous fitness genes together. What the Visitors described does not seem to be asexual reproduction at all but rather a hermaphrodite body model that has both sexual organs like earthworms here but without the ability also to self-fertilize like earthworms. They mentioned something about reshuffling genes between the two hermaphroditic partners during reproduction, so they might also have evolved a mechanism for DNA pairing similar to what we have. It's fascinating. And we don't even know if they use DNA. But it must be a molecule that is very similar or at the very least has the ability to carry information and replicate it. I would love to know more about that. They gave out a lot of information, but they also kept their cards pretty close to their chests."

"Strictly speaking, I'm not sure they have chests," Sandra observed.

That got a laugh from everyone.

"You know, Jim," Gerry spoke up, "I know something about physics, but I haven't spent much time thinking about your field. Sure, I've heard about DNA, but I don't see how a molecule can carry all the information needed to build something like us."

Sandra showed pretend indignation and said, "Gerry, I've explained all that to you before. Surely you're not admitting that you weren't paying attention!"

"Oh I was paying attention," he said. "Just the same way you pay attention to me when I start telling you about the dynamics of mass accretion around black holes."

After the laughter settled down again, Jim said, "Well, Gerry, it's definitely something worth being curious about." He then walked Gerry and the others through a nice five-minute summary that captured the critical points so well, that even Gerry was impressed.

Gerry said, "Did you say that each of our cells has enough DNA in its nucleus that if the pieces from the different chromosomes were stretched out and lined up end to end, they would be two meters long?"

"Yup, just a bit more than two meters. One meter comes from your father and the other from your mother, and both are in each cell's nucleus. Not all of it codes for proteins, but there isn't as much of so-called 'junk DNA' as they used to think. Some sequences carry signals for turning genes on and off and when and under what circumstances to do so. There are other sequences for maintaining proper structure for coiling and pairing—and other vital functions too. And then there are other conserved sequences where we still haven't worked out their function. And remember, the DNA double strand is a very, very fine gossamer thread—only about twenty angstroms across. You know, an angstrom, about the diameter of a single hydrogen atom. Just imagine; a diaphanous thread the width of twenty hydrogen atoms lined up and longer than you are tall—all in the microscopic nucleus of a single microscopic cell."

Gerry was wrapping his mind around that when Jim went on, "Hey, Gerry, you're in astronomy—you'll like this bit. I used to point this out as a neat, gee-whiz science fact to one of my classes. First off, how many cells are in your body? It's tricky to work out a definitive answer, but reasonable estimates for the number of human cells in our bodies run between 30 and 35 trillion. Let's stay on the smaller estimate so we won't be guilty of exaggeration. OK, couldn't be simpler from here. For the nuclear DNA in your body, 30 trillion times two meters equals 60 trillion meters which equals 60 billion kilometers. I looked up a few things when I first thought of this—it's around 4.5 billion kilometers from the Sun to Neptune, our farthest planet. If you took all of your body's nuclear DNA and strung it out end to end, it would reach from the Sun 13.3 times farther out than the planet Neptune orbits. Heck, that's a bit over one-fourth of the way out to where the Visitors are parked!"

Gerry whistled and said, "That just blows my mind! How deliciously geeky—I love it!" Gerry looked upward running some figures in his head. "Holy cow!" he said. "That means that the nuclear DNA from only 670 people would reach from the Sun to our nearest neighboring star!"

All needed to stop and think about that.

Jim's mind was instantly arrested by a vision of the DNA from all the living cells in Earth's thin surface biofilm, all the animals, plants, fungi and microbes, adding up to form a whispery, ghostly web across the whole breadth of the galaxy—*Earth's signature. Oh the astounding wonders of life.*

Then Gerry asked, "What do you think about the fact that they're from a smaller star than ours? Since it would have a longer life span than our Sun and likely a lower UV output as well, genes on the Visitors' planet might have a lower mutation rate compared to Earth. But life there would also have more time to develop and evolve too."

"Great points, Gerry. They could be ancient life forms that have needed greater spans of time to develop." Jim frowned and said, "It drives me crazy to be tantalized by all this partial information that begs for more detail. How I would love to be able to sit down and talk with them."

Gerry laughed. "You might need a space suit. They use oxygen, but who knows what the rest of their atmosphere is like. I know what you mean though; I feel just the same. I have a million questions for them."

Sandra and Ellen had been trying to break in.

"OK, well and good," Sandra said, "but let's get back to what they talked about last night—they described their pair bonding and their loving and supportive relationships. Is that too difficult a topic for you to talk about?"

"Ha, there it is!" Ellen chuckled. "If we all were hermaphrodites, we wouldn't have these girl-guy problems."

"And maybe not as much fun either!" Gerry winked, and they all laughed. Gerry went on, "But as you all know, I'm a very serious guy."

That brought louder laughter.

"OK, OK—maybe not all the time, but I can be. And I'm into pair-bonding." He reached out and put his hand on Sandra's.

She pushed him away but with a smile.

"Well," Ellen said, "I think telling us of their loving and supportive relationships is one of the most important aspects revealed last night. They might have developed that pair-bonding trait long ago, but they were talking about it in the present tense too. It's very important if they have retained it through all their changes and even refined it. My point is that now they seem to be rational creatures who talk about how they value peace and compassion and cooperation and how they have long-term loving pair bonds. If true, they have a lot in common with what we think of as our best features. It ought to be possible for us to make a genuine emotional connection with them. Of course I'm assuming that what they have told us about themselves is the truth. If it's not, then nothing we discuss or conclude has any point."

"That's good, Ellen," Jim said. "I think you've wrapped it up really well."

Gerry needed to head to the university, so they said their goodbyes and agreed to meet at 6:30 a.m. the next morning in Ellen and Jim's kitchen for breakfast.

After they left, Ellen said to Jim, "It's sad to think there are only a few more mornings like this."

Chapter 9

UNEXPECTED

Washington, DC, Saturday, 27 January, 7:00 a.m.

President Kaitland and her chief of staff had settled to listen to the aliens' fifth set of answers. As usual on the hour, the fourth set stopped and a new message began. By the time the first sentence had finished, listeners who were still not fully awake felt a surge of adrenalin that jolted them to alertness more than their espresso. Rachel Kaitland leaned in closer.

> "Dear Friends, we come to you this time with something new—something different—a change of our earlier plan. Five days have passed since we began sending these answers. We have received from you more than thirty-one million questions, and more are continuing to arrive. Some questions were duplicated many thousands of times being asked in different ways and in different languages. We have examined all that we received and have already answered the ones we believe most relevant to introducing ourselves to you. The questions we have already responded to comprised more than 59 percent of the total number received. The remaining 41 percent contain

less duplication and are questions that we are not prepared to answer at this time."

President Kaitland felt the hair on the back of her neck stand up. She and many others felt growing anxiety by this point.

"Of that latter group, two-thirds were questions that pertained to personal issues for individuals or specific agendas for organized groups. Many of those questions concerned issues for which we would be happy to offer help. But they are private issues that should not be broadcast to everyone. The remaining one-third of the questions we have not answered relate to issues of science and technology. Most of these questions came from governments, universities, corporations, and similar institutions. Scattered throughout the unanswered questions were a disappointing number asking us to harm other people and asking us to defeat and even kill perceived enemies. We would never participate in any activity like that.

"We also cannot give you certain technologies in your current war-like state. We cannot give you technologies for which you are not ready. We apologize as previously for what must be seen as a condescending judgment.

"There is another reason beyond the fact that some technologies would be misused or be dangerous to you. If we gave you information that allowed you to leap over so much research and authentic discovery, we would rob you of earning and internalizing your own understanding of such advanced concepts. We believe that it would be harmful to your future progress both in science and in social advancement. Moreover, the majority of our technology would make no sense to you at your present stage. Understanding in some areas must be approached gradually a step at a time. Each step must grow strong in your mind before it can provide support for the next. Some knowledge cannot be rushed. Therefore, we are withholding our answers to those kinds of questions.

"Coming back to that group of questions for which we believe we can help, we request your permission to approach closer to Earth. These kinds of questions can be addressed much better when there is the ability to have multiple follow-up questions and responses—conversations if you will. As we send this request to you now, we know that we cannot have an answer from you before eighteen of your days. We also know that you might need a substantial amount of time to decide upon your answer. We will, therefore, wait until we hear your response.

"In the meantime, we will continue to transmit information that we believe you will find both interesting and useful. By way of explanation, we assumed our present remote position as you began to take significant steps off your planet. Prior to that time when we first arrived in your system, we scouted closer to your planet using autonomous probes to learn more quickly. At times our probes entered your atmosphere and approached quite close, but we always were careful to conceal our presence, and we never interfered with your affairs. We plan to share many of our recordings from those explorations. We surveyed your planetary system examining the major planets, their satellites, and features of your star. We will share much of that information as well. We believe that such information will not prove harmful and might be beneficial in unexpected ways.

"We apologize if this change from the pattern of the last five days is jarring and hope that our explanation is sufficient. We will now begin sharing a series of the records we just described and will begin a new set each twenty-four hours as before."

The alien's gentle voice ceased, and the screen went blank for a few seconds before the message repeated in its entirety seventeen more times.

At the end of those repetitions, the screen went blank again except for the words "Stand By." A few minutes later, a view appeared of the far side of the Moon three-quarters in sunlight and filling perhaps a third of the screen. Beyond it in the blackness and much smaller was a view of Earth also three-quarters lit with sunlight. At the bottom of the screen crawled the words—"First approach to planet of radio source—its large satellite in foreground." A series of unintelligible symbols followed the final word. The view was not a still photo but a video showing the moon slowly growing larger until it passed beneath the camera's field of view.

The image of Earth began to grow until it filled the screen, and the passing scene began to show what must have been the view from an orbit over the equator. Text continued to crawl across the bottom of the changing images updating with notations about the distance from the surface and indications of particular features. The video looked like a modern science-fiction film—someone's imagined idea of a voyage to Earth. But for a huge number—scientists, historians, and any person who grasped the significance of these images—it was overpowering—electrifying to understand that these brilliantly clear images were the first view of Earth by a spacefaring alien civilization in 1916. Many viewers wept, overcome with emotion, including President Kaitland later when she first had the opportunity to watch it alone.

As the images of Earth continued tracking through selected views from both space and from much much closer, it became obvious that this first presentation was an introductory compilation made by the aliens on first arrival. It covered both populated and wilderness areas. It was a mystery how many of the close detailed views could have been obtained. They were far too detailed for some kind of telescopic means, and speculation favored some sort of stealth drone that could approach very close without detection. There were even detailed views of World War I soldiers huddled in misery in muddy trenches. In some views, the faces of individual soldiers could be recognized. Up until these images were being shown all around Earth, there still had been a few people who doubted that the aliens were real because they did not understand the nature of the evidence for the aliens' presence that scientists had accumulated. But now after seeing all of these new images, only those few who believed that the Moon landings were faked or that Earth is flat were still able to believe that the aliens were a hoax.

President Kaitland's analyst team arrived at her usual 0800 meeting while the repetitions of the message were still running. As people took their places at the table, President Kaitland saw that everyone was agitated or excited. She convened the meeting and said, "After the pattern of the previous five days, I am as surprised by the change this morning as I'm sure you are. I would like to hear your first impressions of the announcement."

A clamor erupted as her advisors began talking over each other, and she needed to pull them back to order before she began working around the table. It was soon clear that these analysts who were accustomed to looking for traps, were alarmed at the sudden change. They didn't believe that there were not more general questions the aliens could have answered but instead thought the aliens did not want to reveal more about themselves. More than a few seemed angered by the aliens' condescending judgment of humanity and even angered by their apology for that judgment. The categorical rejection of sharing advanced knowledge with Earth rankled the majority. But the greatest bombshell for all was the formal request to approach closer to Earth. On this point, opinion over how to answer was divided like before when General Beckworth first brought it up.

President Kaitland was still waiting for more reasoned responses and hadn't heard from either General Beckworth or Dr. Schneider yet.

"General Beckworth," she said. "Would you like to share your reaction to the announcement?"

"Thank you, Madam President." Tom Beckworth had been looking down at his notes. Now he looked directly at her and said, "I believe this morning is a pivotal moment for all of humanity. What I heard an hour ago seemed consistent with my view that the aliens are playing with us to see how we will react to new stimuli that they provide. We are the objects of scientific study. They will give us lovely pictures of our Solar System but no more information about themselves or their home civilization. And now they ask to come closer to us purportedly to be more helpful.

"We have already discussed the advantages of being able to carry on repeated close conversations with them. I'm sure many here would use the opportunity to try to persuade them to change their minds about sharing their technology. And I'm sure our agencies would pursue every chance to extract more information to shed light on their true intentions. But if we agree and they do come closer, we don't know whether they will approach close enough to be in reach of our defenses. That must be considered if we do decide to agree to their request.

"And if we refuse their request, what then? Perhaps they will depart; perhaps they will approach without our permission. We don't know. If they depart, we will remain in this position of not knowing what to expect from them in the future or when to expect it. We would only know that we would need to be single-minded in devoting ourselves for hundreds of years to advance our capabilities and technology as much as possible to be prepared for their inevitable return. It would be a life of preparing for war for centuries to come.

"Finally, there is another point we have not addressed. If their technology is as advanced as it appears, they could be recording this meeting and relaying a signal to their vessel. We don't know what they might have planted here before they moved to their present position. We have heard how they can put themselves into different machines they call their bodies. I don't want to sound too extreme, but we must consider it plausible that some could be walking among us in bodies indistinguishable to us from humans in normal circumstances. If we are to have any hope of improving our odds, we should invite them to come closer—close enough both to talk and to be in range of our missiles."

President Kaitland was not surprised at Beckworth's comments. She could see that he had hardened his attitude toward the aliens and was concerned that he might be losing his perspective. Still, he had offered the most substantive analysis she had heard so far. "Thank you, General Beckworth. Let's hope that the aliens are not, in fact, recording this meeting and others or what we decide to do will have little impact." She turned toward Jonathan Schneider and said, "Dr. Schneider, would you give us your impressions please?"

"Thank you, Madam President," he said. "I would first like to express an observation to some of the earlier comments that showed resentment about the aliens' judgment declaring that humans are not socially ready for their advanced technologies. It's worth remembering that we do not offer small children matches to play with, and we do not sell cases of dynamite to teenage boys at the corner hardware store. I did not take any offense from their statement and can easily see that it was appropriate from their point of view.

"As to the rest, I do not see evidence that the aliens are 'playing' with us as General Beckworth suggested, but I do agree with him that they might be testing our responses. They have already had so much time and opportunity to analyze our behavior that I think they know us well. But it is possible that this request to come closer is a test—a test of where we are in our ability to look past threat and reflexive defense responses and instead consider the advantages of being open to trust and the possibilities of what could come from cooperation. A rejection to this request could close the door to them for centuries or our agreement could open the door to a very advantageous relationship. In that sense I agree with General Beckworth that this could be a pivotal moment in human history. I too favor inviting them to come closer, but only for better communication. We then could begin to build the basis of a relationship. I think they would see through any attempt to entice them to come so close that they would be in range of our warheads. Close enough to converse is good enough."

Again Rachel Kaitland found Schneider's points to be sound. She thanked everyone and asked them as usual to have their refined reports to her by 1800. She left for another meeting, anticipating a busy day ahead—a day where the whole planet was reacting to a new and very disturbing problem. She had already asked her secretary to schedule calls to the leaders of a half dozen of Earth's most powerful nations.

The first of the aliens' video records took almost eight hours to complete before it began to repeat. As the morning in the West Wing ended, the first presentation was still in progress. Every media outlet was erupting with images taken from it. Experts discussed the recordings' significance and government officials were questioned about the aliens' request to come closer. Spot poll results indicated mixed feelings among the populace ranging between awe at the spectacular images and anxiety over the aliens' request. President Kaitland knew that the aliens' request would be the consuming issue for the foreseeable future.

Canberra, Sunday, 28 January, 7:00 a.m.

By midafternoon in Washington and morning in Canberra, the four friends had gathered at Jim and Ellen's kitchen table for breakfast. Three more mornings now remained before the two Americans would be on their way back to Arizona.

Gerry, bolting a cup of coffee, was very excited despite having stayed up half the night watching the new historical video. "It's just one amazing thing after another with these guys," he said. "If this first sample is typical, they may have tremendously precious images of historic events from around the world since 1916. And the quality of the images is astounding. Our regular TV screens display a very good resolution these days. But I was able to get in touch with a friend at MIT in the middle of the night who told me that there is way more information in the signal. His group had been able to run the transforms on the signal and found the extra information embedded within it—an eight-fold higher image resolution! Think of it, eight-fold greater detail than our current ultra-high definition standard. And not only that, the signal includes image data for wavelengths we can't see! My friend said they hadn't finished working all that out yet. They will have to transform the extra signals and feed them into conversion displays to get an idea of what is captured. The Visitors told us before that they could see into the infrared wavelengths beyond what we can see. They must make their cameras standard with the same capability they evolved with plus more into the UV end too that I suppose their new machine bodies can handle. We'll have to play with how we interpret the signal in order to find all the information in it."

Jim was laughing. "Well Gerry, I'm glad to hear that you found last night's broadcast somewhat interesting. Any possibility you'll let the rest of us talk too?"

Everyone laughed while Gerry shouted, "Probably not!"

Jim was in good spirits and found the change in the Visitors' transmissions to be a welcome one. Gerry was over the moon about the aliens' promise of solar system data. He said, "It could be months before Earth decides how to answer the Visitors' request. If the transmissions go on for that long, the amount of scientific data collected could add up to more than humans would be able to gather on their own over the next two centuries."

"I wonder what they could mean about helping with millions of questions of a private nature?" Sandra mused. "It doesn't make sense. And even if they could fly down and come to tea, what kind of private matters could they help with—what would they know about things like that?"

Ellen laughed and said, "Hey, maybe they have zmail addresses for everyone. Oh wait! Yes—maybe they have board-certified interstellar psychological counselors with them too." More laughter.

Jim looked around and saw that all had relaxed into a less-fearful attitude about the aliens' true intentions. For the last several days, the gentle alien voice had said all the right kinds of words to appeal to people such as these four. It all sounded so good to them that their desire to believe was overcoming their anxiety.

Gerry was about to say his goodbyes and leave for work when a firm knock sounded from the front door. Jim and Ellen looked at each other.

"Were you expecting anyone?" Jim asked Ellen.

She shook her head no.

The knock was repeated, and Jim stood and walked to the front door. When he opened it, he stiffened in surprise. Two men dressed in suits flanked the doorway and just behind them stood a large uniformed Australian federal officer wearing a sidearm. All three wore sober expressions.

Jim furrowed his brow and said, "May I help you?"

One of the suited men said, "Is this the residence of Dr. Ellen Hazelstein? Is she here?"

Jim's brow furrowed more—he noticed the man had an American accent. He said, "May I ask what this is about?"

"Please just answer the question, sir," said the other man.

Jim thought fast but saw no safe alternative. "Yes, she is here," he said. "I'm her husband. Please tell me what this is about."

"May we come in?" the second man said.

Ellen, Sandra, and Gerry had left the table and were watching from the kitchen doorway.

Ellen said, "I'm Ellen Hazelstein. What is this about?"

"May we come in?" the man repeated. "We're from the American embassy. I'm Officer Stanwell and this is Officer Lee." Both men held up their embassy photo ID cards.

Ellen's eyes widened, and she said, "Yes, of course." Her mind raced—*where did I put my passport? Is there something wrong with it?*

The men came in and politely asked Jim to close the door.

Officer Lee said to Ellen, "We must first verify your identity."

Ellen by then remembered where she put her passport and retrieved it for him.

After he examined it and compared her with the photo, he handed it back and said, "We have a confidential matter to discuss with you. Is there a private room where we can talk?"

"No," said Ellen too quickly remembering the bedroom's present messy state. "This apartment is tiny. I have nothing to hide from my husband or my two close friends here. Is there something wrong with my passport?" Her face showed worry.

The suited men looked at each other, and Officer Stanwell nodded.

Officer Lee turned back to them and said, "We received a call two hours ago from the Office of the President of the United States. President Kaitland has directed us to find you and tell you that she requests your presence in Washington for a meeting with her as soon as possible. We are aware that you are already scheduled to return to Arizona on 31 January. Instead we have arranged a flight for you tomorrow that will take you to Dallas, Texas, and another flight from there to Washington, DC. Arrangements for your hotel near the White House have already been made. You will be met at Dulles International Airport by White House aides who will take you to your hotel and provide you with further information about your stay and your meeting with the president. The US Government will pay all expenses including for your flight to Tucson, Arizona, after your meeting with the president is finished."

Ellen, Jim, Sandra, and Gerry stood there staring with their mouths open and their minds uncomprehending.

Chapter 10

DECISIONS

Washington, DC, Saturday, 27 January, 4:00 p.m.

General Tom Beckworth was troubled—troubled and frustrated. *The others are not taking the threat of the aliens seriously enough. The danger is so obvious! Why aren't the others pushing to begin building effective defenses as soon as possible?*

Even after the latest message from the aliens and their request to come closer, too many still wanted to take a wait-and-see approach. That seemed foolhardy to him. Sometimes his mind stumbled into brambles of worry—desperate worries that pierced and would not release him. Yet the same facts did not seem to provoke the same intense feeling in others. Still, he knew that these same people were rational, intelligent, and experienced analysts. Sometimes his reason could outshout the worry and remind him of that. He had begun to wonder if he was missing something.

He decided to call Joe, his oldest and most trusted friend. He knew of no one who would have a more clear-headed view. Saturday afternoon, he spoke Joe's name into his phone. After a few rings, Joe answered, and Tom said, "Joe—great! I wasn't sure I'd catch you. I'm guessing you've been as busy as I have."

"Hi Tom! Hey, it's good to hear from you," Joe answered. "It's been too long—I can't believe what's happened since you came over for dinner that Saturday before New Year's Eve. Yeah, it's been crazy-busy here too."

"A month! Hard to believe," Tom said. "After these insane last weeks, I've lost track of the time. How's Tara?"

"Tara's great. She's been down in South Carolina this last week visiting her mother and won't be back until Wednesday. Hey, since I'm batching it for now and if you're not tied up, why don't you drop over this evening for a good catch-up? I can order pizza and we can have a bull-session like the old days."

"Perfect! You won't believe this," Tom said, "but I phoned just now to ask if you might have some time soon for a private talk about some work things that have been on my mind. That OK with you too?"

"Sure," Joe answered. "We can talk as late as we need to and solve all the world's problems. I plan to escape this asylum early today. I'll be home by 6:30. Come on over then."

"Great!" Tom said. "See you then."

Joe and Tara had a pleasant home in Alexandria, Virginia, not too far from the Pentagon. They had hosted Tom to dinner and barbeques many times. And for the last ten years, Joe and Tara had gone out of their way to look after Tom since his wife, Charlotte, who was also their friend, had passed away from cancer. By seven, Joe and Tom were settled in Joe's family room with drinks in hand and a hot pizza just delivered.

Both Tom and Joe had been assigned to the Pentagon in Washington for the last two years after spending most of their careers in the usual nomadic life of military people. Both had been extremely bright lads. They were roommates on full scholarships at the Air Force Academy and graduated in the same class. They were thrilled with the idea of joining the still new Space Force. Both were very serious about their studies and loaded their schedules with mathematics, physics, and engineering courses that might give them a leg up. After graduation they managed to cross paths numerous times including an early period when both were assigned to a unit of the new Ground-Based Strategic Deterrent system—the new ICBMs that replaced the old Minuteman III missiles. The Air Force, not the Space Force, controlled all of it, but they were seconded temporarily as recent Academy graduates to gain experience in missile systems.

Although their paths diverged later, they still met often on various trips to Washington or at meetings with defense contractors. Joe's talents leaned a little more toward technical engineering issues, and Tom's talents tended more toward military strategic thinking. By the beginning of their fifties, Tom found himself with a one-step higher rank than Joe, and both were in the prime of their careers and well respected in their specialties.

Joe said, "Well, what's up with the work things you mentioned? Must be about the aliens, right?"

Tom sighed and said, "Yes, no surprise there."

Tom knew the rules against discussing sensitive information outside of meetings like the ones he was involved with at the White House. But both Tom and Joe had top secret clearances, and Tom knew Joe was already involved with his own analysis teams working on defense capabilities related to the alien issue. Tom outlined all of his concerns, the same ones he had expressed to President Kaitland at the meetings of the last week. He looked Joe in the eyes, his face and tone earnest. "Joe, the entire future of humanity might be at stake in the next few months. This is like nothing in human history. This could be the end of the human story on Earth! Why doesn't everyone see that?"

Joe saw that Tom was under much more strain than he had realized—that Tom was assuming too much of the weight of the problem onto his own shoulders. But he continued listening.

Tom said, "We're lucky that Kaitland is president. With her brains and background, we can rely on her to make rational decisions. But I'm worried about the influence of the people lobbying her to wait and let events play out longer before we *do* anything. We need to do so much that even a few weeks delay could make the difference for our survival. We both know all of our existing nuclear missiles are designed only to reach other places on Earth's surface and wouldn't work for targeting high Earth orbit regions. It would take a massive concerted effort to put together nuclear armed missiles that could fight in space. If a plan is approved soon, we might have a chance to deploy some in time depending on how much time the aliens give us. You must have been identifying resources for that already."

"Yes. It's all highly classified, of course," Joe answered. "You're right; a plan like this would require a greater flood of human and material resources than we've ever mobilized before. Assuming Congress agrees and funds it, it'll take a miracle to have everything come together fast enough."

"What's that old saying—the prospect of being hanged in the morning focuses the mind wonderfully? Well, a threat like this might provide the miracle we need. I want to believe we'll give ourselves the chance to get through this. But the discussion and debate just goes on and on. Too many don't want to take a decisive stand. We can't wait to know what the aliens are planning. What if they're preparing to launch an assault right now? We could be caught with no way to defend ourselves—just fish in a barrel," Tom said.

This was the moment when Joe knew his friend was stumbling under the weight of the scenarios in his mind. He had noticed Tom's haggard appearance. *When did he last have a solid night of sleep? Tom is an extremely rational and clear-thinking person. This talk is way out of character.*

Joe said softly, "Tom, please listen to yourself."

Tom started to say something but then stopped and slumped looking very tired—exhausted.

Joe looked at Tom and continued, "Tom, you know that fatigue and worry cloud the mind. Strain distorts your view. All of them together for long enough can turn you upside down. Remember how you were after Charlotte died. For a long while you couldn't recall large parts of your normal work expertise. You could only remember all the things you and Charlotte had said to each other—all the things you and she had done together, places you had gone together. I think it's a little like that now even though the reasons are very different. You look completely exhausted. I can see the strain in your face."

Tom remained silent, slumped and looking down.

"You can't assume the whole burden of saving humanity onto yourself. You need to trust other people to do their part too. There are tens of thousands of us working on this problem now. It's been less than a week since we began hearing answers from the aliens. Thousands of us have already begun taking action. From where I'm standing, planning and searching for solutions have never moved so urgently. I'm confident that a major decision will be coming soon."

Tom sat quietly looking down while Joe waited patiently.

Tom knew he could never hope for a truer friend than Joe. Being here now, hearing the words of his trusted friend—words that made sense—brought on a calmness he had not felt in weeks. He could hear the reasonable part of his own mind talking to him again. It was like waking up from a bad nightmare when you realize you are not in the terrible place your dream had taken you. Relief began to wash over him, and his mind stepped out of that dark labyrinth.

At last he looked up and said, "Thank you, Joe—more than I can say. I guess I felt lost—that this might be the end of humanity's time, and maybe no one else would be willing to step up." He paused and sighed. Then he took a deep breath and said, "I feel a little foolish. Thank you, Joe, for pulling me out of the hole."

"Don't mention it," Joe said. "You would do the same for me."

They sat for a few minutes saying nothing and sipping their drinks. Finally Tom said, "Well then, let's do what we need to do first anyway. I haven't been as closely involved in the hardware side the way you have. Can you bring me up to date?"

"Sure," said Joe. "The Outer Space Treaty of 1967 has been under a lot of pressure lately, but it's still in force so nothing has been built to deliver nuclear warheads that far out. We would need the biggest crash development program imaginable. Four commercial companies already have lunar-capable rockets operational and have been jostling for Space Force contracts."

"And I'm sure you've already thought of the newest private rocket waiting in Earth orbit to be fueled for launch to Mars," Tom said.

"Right," Joe said. "That's where I was heading next. There are a lot of barriers to get through because of the treaty, but if we could get past the approval problem, it should be possible to reconfigure the rockets. We're just military though. All of that is a political process. But an emergency like this might be enough to push it through."

They talked well into the night. It helped to think about things they understood and which they might be able to influence. By the time Tom left for home, he felt better than he had in weeks. Things felt on an even keel again.

And Joe? Joe felt enormous relief to see his friend back to his old self.

Canberra, Sunday, 28 January, 9:00 a.m.

Ellen and Jim needed a moment to recover from the American embassy message. When they had regained their composure, Ellen asked the two men why she was being called to Washington, but the embassy men had no more information for her.

Ellen asked, "Does the arranged flight include Jim?"

"Just you, ma'am," Officer Lee said.

"How long will I need to stay in Washington?" she asked.

Officer Stanwell said, "We don't know, ma'am. We'll send a car tomorrow morning to take you to the Canberra Airport. It will arrive at 8:00 a.m. Please be ready on time."

Officer Lee seemed a little more sympathetic and said he would inquire after they returned to the embassy to see whether any more information might be available for her.

DECISIONS

They left.

Ellen sat down on the sofa and looked at the other three, her face worried.

Jim sat down and put his arm around her. He said, "Don't worry, sweetheart. There must be a reasonable explanation for this. It must be connected somehow with that commission panel you sat on a couple of years ago. It must have something to do with the aliens. I mean what else could it be?"

"Yes," she said. "The same idea came to me. But I can't imagine why they would want me particularly and so quickly too."

"Maybe they're asking all the panel members back," said Jim looking doubtful. "But you might not need to stay long. They mentioned they would arrange to get you back to Tucson after your meeting with the president. You might even get home about the same time I do," he said hopefully.

Sandra and Gerry left saying they would check back with them later.

Ellen began gathering her things together to pack. She was glad they both had a habit of packing little even for longer trips.

Later that afternoon, there was another knock at the door. The same two men from the embassy had returned but without the uniformed officer. Officer Lee said, "We know it was a shock for you this morning. We checked through our secure channel to Washington and were able to get a little more information for you." He handed her a sealed envelope, thanked her for her cooperation, and both left.

The envelope contained a short, printed zmail message that read, "I apologize for this interruption to your trip, but there is some urgency. I believe we won't need to keep you in Washington more than a day. Thank you for your help." It was signed R.K.

By that evening Ellen was finished packing, and they relaxed with a walk at dusk on the nearby park-like university grounds. They loved listening to the liquid warbling serenade of scores of magpies perched on the ramparts of university buildings. Jim said, "I hope we'll hear an argument between a couple of kookaburras too. They always make me laugh." But the kookaburras remained silent.

Jim and Ellen felt sad that this was their last walk together in Australia.

That night they began watching the next installment of videos from the Visitors—the new images were just as compelling as the first set and were still dealing just with Earth. But Ellen and Jim were too tired and distracted to appreciate them fully. After a half hour, they turned off the TV to go to bed.

Earlier in the afternoon they had canceled their standing plan of breakfast with Gerry and Sandra. Their plans were on 'fast forward' now.

The next morning Jim assured Ellen he would take care of everything that still needed to be wrapped up in Canberra. "Don't worry," he said. "I'll see you again in a couple of days. I know that whatever it is they want from you, you'll do brilliantly."

Before they knew it, the clock showed 8:00 a.m., and the embassy car pulled up outside. Jim carried her bag to the car, they embraced warmly ... and then he was alone.

✦

The second set of videos also ran for eight hours before beginning to repeat. All around the world, government experts, university and private research groups, historians, and media analysts were wedded to their screens and instruments to extract every possible bit of information from the broadcast. It was a motherlode of treasure for historians. Unlike the first offering that had mysterious symbols after the text identifying the scene, the latest set showed the time, date, and location of the scene along with the explanatory notes. Speculation was that the aliens had overlooked translating the times from their own system to Earth-style in the first set. In any case, knowing the time and date of the scene was crucial.

Many scenes from space shown in extended wavelengths provided valuable weather data and cloud patterns for periods after 1916. That data from early in the twentieth century would better validate existing weather forecasting models. And then there were fascinating glimpses of known events like watching the shadow of the Moon march all the way across North America from the total solar eclipse of 1918. Sometimes close-up scenes of historical importance were accidentally swept up in the alien surveys. There were views of crowds surging into streets in London, Paris, and New York City to celebrate the end of World War One—all such events clear from the accompanying locations, dates, and times shown at the bottom of the images. Based on the first two sets of recordings, the aliens were working through their early survey records of Earth first. Planetary specialists were eager to see the studies of the rest of the Solar System too. Most watchers were caught up in the excitement that helped them forget to worry about what might happen next.

But that was not the case for governments anywhere in the world.

◆

Washington, DC, Monday, 29 January, 8:00 a.m.

In Washington, the question of what to do about the aliens' request to approach closer to Earth was the new center of every strategy meeting. President Kaitland could not focus on the latest alien transmissions that began every day—as fascinating as they were—she left that to others. She was working with two groups. The first was analyzing how to proceed in dealings with the aliens if Earth refused their request—the second was planning how to deal with the aliens if Earth *did* invite them to come closer. The defense and intelligence analysts she had been meeting with every morning comprised the second group. She met with them as usual on Saturday and Sunday, and by today, Monday morning, she hoped to consolidate all they had discussed. She still hoped for an inspired contribution at one of her meetings to help her decide on the best course.

The day before, on Sunday, they clarified what relevant defensive capabilities existed and were available. A critical question was how much time they would have to expand those capabilities. If they invited the aliens to come closer, everyone expected they would need at least several months to make that journey. No one expected them to accelerate to interstellar cruise speeds within the Solar System where random clumps of matter still orbit the Sun. Passing through the Kuiper Belt at that speed seemed especially hazardous to the analysts. So that added a few months to the time Earth would have to organize and strengthen defenses.

But even more important, President Kaitland expected it to be difficult for national governments to reach agreement over the aliens' request. Six months seemed an optimistic guess for that decision to be made. She had a meeting set Monday afternoon with State Department people on that specific issue. Meanwhile, Defense Department officials were thankful that they had around nine months to prepare for the aliens' arrival.

President Kaitland felt optimistic about the day. Things were coming together well. She had even been pleased to notice Sunday morning that General Beckworth was more balanced and constructive than at previous meetings. She convened the meeting and began to poll the group asking, "Where should we ask the aliens to position their vessel when they arrive near Earth? How close do we want them?"

Opinions varied over a range as far as halfway to the orbit of Mars to a position near the orbit of the Moon. Discussion made clear that different solar orbit positions such as one halfway to Mars, would be a problem. Soon Earth and the alien ship would find themselves far apart—in time their two orbits would put them on opposite sides of the Sun—too great a distance for better communication and certainly for defense strategies. Even a position at one of Earth's two closest

Lagrange points, L1 and L2, that would keep the alien vessel linked with Earth, would be too distant for defense purposes.

An Earth orbit similar to the distance to the Moon garnered the bulk of the votes. The ship would stay with Earth and be about 1.3 light-seconds away. But General Beckworth, with a few others, objected. He said, "Madam President, at a distance as far as the Moon, Earth's ability to project force is still too limited. In case a situation arose that demanded action, Earth would have no reasonable way to fight back."

President Kaitland asked, "What kind of situation short of an attack are you thinking about?"

"I'm sorry, Madam President, but I do not have an answer to that question," Beckworth replied. "I could imagine various events based on our interactions with other humans, but if a situation requiring us to fight the aliens did arise, it could be over something new and unanticipated. Everything about this situation is mired in a lack of critical information concerning every key feature—things like how fast the alien vessel can maneuver in close quarters, what range could it attack from if they have lied about not having weapons, how devastating might their weapons or 'tools' be and what would we need to watch for to perceive that an attack might be imminent. The only way I know how to deal with a situation like that is to box an opponent in to a smaller space and have the ability to attack that space from more than one direction quickly and simultaneously. My recommendation now is to invite the aliens to come much closer and enter a high Earth orbit where we would already have stationed a number of defensive weapons in readiness. I would place as many armed missiles as we can gather into strategically located high Earth orbits and have them positioned so that at least several could be rapidly deployed to whichever point the aliens station themselves."

This was a new and more ambitious suggestion than had come up before, and it caught President Kaitland a little by surprise. She noticed murmurs of approval around the table.

She said, "You realize that the Outer Space Treaty forbids any such plan."

"Yes, Madam President, I do," Beckworth answered. "But the treaty was confirmed eighty-five years ago to prevent humans from attacking other humans from space with weapons of mass destruction. This situation, while violating the letter of the treaty, would not violate its intent. I know it is outside my purview to suggest it, but I think that the signatories to the treaty might agree to such a plan, might also agree to participate with some of their own resources, and in particular might agree to a specific waiver of the treaty for a unique situation like this."

President Kaitland paused, thinking. Then she said, "Thank you, General Beckworth. Let's discuss this idea more."

After another hour, there was clear agreement that the concept should be given serious analysis. She knew that to implement such a plan would take every bit of time they might have. She asked General Beckworth to write a more detailed description of his concept to be ready by tomorrow's meeting.

He was delighted with her request! It felt as if his call to arms might be heeded at last.

Chapter 11

HISTORY LESSON

Washington, DC, Monday, 29 January, 6:00 p.m.

A quirky fact of traveling from west to east across the International Date Line is that when flying from eastern Australia to the US, one lands in the US at about the same time of day and on the very same date as when the plane left the ground in Australia. If traveling only to the US west coast, the plane often lands there "before" it takes off in Australia, an amusing oddity for travelers to joke about. In Ellen's case, she landed at Dallas-Fort Worth International Airport at almost the same time and on the same date as she left Sydney.

By the time she completed her next flight and landed at Washington Dulles International Airport, it was evening Washington time. She made her way to the baggage collection area and scanned her baggage ticket at one of the many scanners. It directed her to a particular pickup station where she flashed her baggage ticket again. In thirty seconds her bag slid down a chute and onto the tray in front of her.

She thought of Jim, now so far away. She was eager to get to the hotel so she could settle into a quiet room and phone him. Though the US government had splurged on a business class seat for her, she had a great deal on her mind and managed only an hour of light sleep during the long passage over the Pacific. Modern planes were more comfortable than they had ever been but were still subsonic. Her flight from Sydney to Dallas took about sixteen hours. A new hypersonic passenger plane service was the one exception to conventional aircraft. It flew above nearly all of the atmosphere at Mach 5.3 and had gone into limited service three years before. But, like the old British and French Concorde, it was limited to a small number of elite travelers with very deep pockets.

Trying to forget her fatigue, she walked toward the exit. *How will I find the people sent to meet me?*

Before she had taken ten steps, she was hailed by two people who approached from nearby and called to her by name. They must have known what she looked like, she thought, and perhaps had the means to know where her suitcase would appear. They introduced themselves as Officer Sarah Whitten and Officer Frederick Sorenson and showed her their White House ID badges identifying them as White House security staff. They welcomed her to Washington and were smiling and chatty as they led her from the terminal to a large, black car parked under a sizeable sign proclaiming "No Parking Zone." Sarah ushered Ellen into the back seat while Frederick put her bag in the back. As they pulled into traffic with Frederick driving, Ellen felt a wave of weariness as the sleepless hours caught up with her.

But Sarah turned in her seat and said, "We know you must be very tired right now, so you might be glad to hear that we have nothing planned for you until tomorrow. We hope you will be able to have a good night's sleep by then. I think you will be very comfortable in the new Marriott, and it's only a short walk from there to the White House for tomorrow. The president is busy all morning but has you scheduled for a meeting over lunch and has also reserved two more hours with you after lunch. That's unheard of—you must be very special."

Ellen snapped back to alertness.

Frederick said, a little sharply, "Sarah."

"Oh." Sarah paused—then continued, "So we will get you settled into the hotel and arrange room service for you. A representative from the President's Office will knock on your door tomorrow at 11:00 a.m. to take you to the White House. We will leave a packet of information for you that will tell you more about that person and about the meeting."

True to their word, they confirmed all the arrangements for her at the front desk, accompanied her to her room, and asked if there was anything else they could do to make her comfortable. "Thank you," Ellen said sincerely. "You've been very kind. And yes, you were right; I'm very tired." Sarah and Frederick smiled and handed her the packet of information as they said goodbye.

Ellen was relieved to stretch out for a few minutes before calling Jim. While the phone was ringing, she opened the manila packet and took out several pages of text and a cover letter on White House stationery. The letter thanked Ellen for traveling so far to be of assistance. As she expected, the letter did not identify the purpose of the meeting or even that it was with the president.

In case it fell into the wrong hands, she thought.

But it did provide the name of the aide who would come for her in the morning—Nancy Birchfield.

Then Jim answered his phone. "Hi, sweetheart! I was glad to see your name on my phone just now! Did your flights go OK? Is everything on track so far?"

"Yes, darling," she said. "I'm glad to hear your voice again. The flight was fine and uneventful—my favorite kind. It all looks good for now. I go to the White House tomorrow morning at eleven, so I can have a good night's sleep first."

Jim explained that he had wrapped up everything for the apartment rental and was almost finished packing. "Gerry said he'd pick me up tomorrow morning to take me to the airport. We might even get back to Tucson on the same day. Wouldn't that be great!"

Ellen heard a knock on the door. On opening it, she saw that it was room service with her dinner and realized that she was hungry. "Jim," she said, "I'm so glad everything looks good back there for you to be on the way tomorrow. Meanwhile I'm exhausted. I'll have my dinner now and then pass out for the night. I expect jet lag will wake me at an annoying time in the middle of the night, but at least I can get some sleep for a while. Good night favorite husband. Don't forget to board the plane tomorrow!"

After they finished their goodbyes, Ellen looked at her dinner tray and noticed a tag—a vegan dinner. *How perfect!*

She settled down with it and picked up the information papers again. She had hoped for more about the purpose of the meeting but was disappointed to find that the three pages after the short cover letter were photocopied excerpts from one of her books with some of the paragraphs highlighted. As she read them, she guessed what the meeting would be about. The pages referred to what happened to Native American peoples after the arrival of Europeans. As she fell asleep, she tried to remember some of the things she had said at that meeting of the Presidential Commission on Ethics in Government two years ago.

Washington, DC, Tuesday, 30 January

A night's sleep did wonders for Ellen, and she managed more good thinking time the next morning. When the knock on the door came at 11:00 a.m., she felt both excited about the coming adventure and a little anxious over whether she would have anything useful to contribute. *Well, relax and keep a clear head.*

She opened the door and saw a smiling woman of about forty. Nancy Birchfield introduced herself as a White House aide.

Ellen was very pleased to see that Nancy had brought with her an extra woman's overcoat. Someone had been thoughtful enough to realize that Ellen had just arrived from midsummer in Australia. She had heard on the morning news that Washington was in the middle of a cold spell. *How nice of them!*

As they walked to the White House, Nancy explained what to expect. She confirmed Ellen's guess that President Kaitland had remembered her from the Ethics Commission meetings Ellen took part in and that today's meeting topic dealt with the aliens. But Nancy did not know or did not reveal more detail.

As they walked through the biting winter morning air, Nancy said, "Two other people will be present during the lunch and will stay for discussion afterward. I don't know how long the meeting will last though. That will be up to the president."

Ellen was glad that she had participated in the commission meetings or she might have been feeling intimidated by now. Still, she was feeling some anxiety. The commission meetings had not been held in the White House, and President Kaitland had attended just one session for about three hours on one of the meeting days. There had been eleven other people on the commission, and Ellen was surprised that the president had remembered her. They entered the lobby of the West Wing and passed through a thorough security check. Ellen felt like a tourist as she was led into the innards of the historic building. Nancy pointed out the famous rooms and features as they walked. She mentioned lunch would be in the private dining room just off the Oval Office and after lunch the meeting would be in the Oval Office itself.

The Oval Office! Ellen felt a significant bump in her anxiety level and had to tell herself to relax and calm down.

Nancy led Ellen into the dining room and asked her to sit down and make herself comfortable. Nancy left to check that everything was still on schedule. Ellen sat alone looking around the room and thinking about all the historic events that happened there. In the next few months, perhaps the most pivotal moments in American history, indeed, in recorded human history, could very well take place in this building.

Nancy returned with the president's personal secretary, Eleanor Atwood, who explained that there was a small change in the plan. She said that President Kaitland was concerned she might need to leave the meeting with Ellen earlier than first planned. The lunch was an informal one anyway, so the president asked that lunch be brought to the Oval Office, and they would eat there and begin the meeting with lunch.

Eleanor thanked Nancy for her help and led Ellen into the Oval Office. She opened the door and said, "Madam President, Dr. Ellen Hazelstein is here for your meeting with her."

Ellen felt a wave of nervousness and was a little starstruck as she first gazed at the president who was standing to meet her. President Kaitland was an athletic-looking woman appearing younger than her fifty-two years and still without gray hair after three years as president. Ellen had seen the president many times in news videos, but in person it was just that one time for the brief period at the commission meeting. It felt very special to be present with her in the Oval Office.

The president crossed the room to them smiling and held out her hand. As they shook hands, President Kaitland said, "Thank you so much Dr. Hazelstein for coming such a long way to meet with me. Welcome to the White House. Please know that I am truly glad you were able to come. It was good that you already had the necessary security clearances from the earlier meeting and that updating them did not take long. Otherwise I could not have invited you here so quickly."

Ellen relaxed. She knew that the president was saying what would be expected, but she sensed genuine sincerity. "Thank you, Madam President," she said. "I am deeply honored that you have invited me. I hope I will be able to be of some help to you."

The president turned and beckoned to a distinguished looking gray-haired man appearing to be in his mid-sixties who came forward extending his hand. "Let me introduce you to one of our most talented analysts," she said—"Dr. Jonathan Schneider." Jonathan Schneider shook her hand while welcoming her and then stood back while the president introduced Allen, one of her aides who would take notes of their conversation.

Sandwiches and drinks had been brought in, and the president invited everyone to sit down and begin lunch. President Kaitland said, "Thank you again, Dr. Hazelstein, for coming so quickly. I expect you've guessed why I asked you here from your book excerpts I sent. We are facing some very difficult decisions now, and I am looking for insights from all directions—from anyone who might be able to help."

"Thank you, President Kaitland. I hope I will be able to help," Ellen said softly. *I'm glad for all the conversations I had in Australia with Jim, Sandra, and Gerry*, she thought. They had helped her think more intensely about what the aliens' arrival might mean for humanity.

President Kaitland continued, "I imagine you have followed all of the alien broadcasts and have heard of their request to approach Earth. Our answer to them might make the difference between a happy or an unhappy future. From the ethics meetings you took part in a couple of years ago and from what I saw in your book, I know that you have thought deeply about the outcomes of clashes between a powerful culture possessing advanced technology and a less powerful group without that technology. I've asked Dr. Schneider to join us because I'd like him to hear what you have to say. He might stimulate more discussion with his own ideas too.

"Even though we don't know how much is true of what the aliens have told us about themselves, we can conclude that their science and technology is advanced well beyond ours. That makes our situation analogous to what you have described in your book, and in this case we are the ones who find ourselves at a disadvantage. Would you tell us about some of the culture clashes you have studied? Oh, and please do go ahead with your lunch. We can talk around the munching."

Ellen had examined this topic often with her classes and summarized the history of such clashes for the president. She explained that the majority of cases by far had resulted in a weakening of or even extinction of the culture having the less-advanced technology. In earlier times in some cases, it resulted in the extinction of the less advanced people themselves. In more recent times, clear examples occurred in the aftermath of the European exploration and colonization of the Americas, Africa, southern Asia, and Australia. Ellen described a few of the more egregious and horrifying examples adding, "I am using the terms 'less advanced and more advanced' as a kind of shorthand referring to the level of technology and the degree of social organization that enables greater or lesser military power. For other cultural features, it could be argued that some of the conquered groups possessed cultural refinements that were more advanced than the conquering culture."

President Kaitland asked, "And were there no stories that had happier endings?"

"Yes, there were," said Ellen. "But not many, and the cultures of the less-advanced societies were usually absorbed into the dominant culture during a difficult period of transition. But some aspects of the absorbed cultures enriched the dominant cultures with more variety, and the survivors from the less-advanced cultures could be viewed as having gained a life with more options than they had before. A great deal depends on how open and receptive the dominant culture is to newcomers. If over time the newcomers are accepted as equals with equal rights, then the clash can have a happy ending. Of course it will always happen that many in the conquered group will suffer great emotional pain over the loss of their previous autonomous way of life."

"Have you had any thoughts about how the future might play out for us in our association with the aliens?" Dr. Schneider asked. "I mean if the aliens are as far advanced technically as they have presented themselves, do you have a feeling of what our chances might be based on how they describe themselves and their society?"

Ellen answered, "Yes, I have thought of little else since the broadcasts began." She summarized her ideas that came out of her discussions in Canberra. She also worked through all the same points that Dr. Schneider and the team of biologists had made—points that tried to project emotional attitudes of the aliens based on their biology and reproduction.

Dr. Schneider and the president glanced at each other surprised and impressed with her comprehensive command of these many issues.

Then Ellen emphasized, "I do not believe we can draw any reliable conclusions from those extrapolations. Please remember, in the case of the Europeans and the Native Americans, both had the same biology. The two groups were members of the same species with all the basic cognitive and psychological features that humans have in common. Not only that, the Europeans professed a religion that at least formally emphasized love, mercy, and compassion. Yet the Europeans were guilty of aggression and grotesque atrocities against the Native Americans. It's true that some of the atrocities were responses to Native Americans' counterattacks to drive out the invaders. But once violence began, it escalated and each side was seen as inhuman by the other. Because of the cultural differences, it was easy for the Europeans to think of themselves as superior and even divinely ordained to subdue the 'savages'. That and their strong desire to gain great wealth and prestige in the New World allowed them to rationalize their cruelty.

"Now think of the aliens," Ellen said. "We have no biology in common with them. And they have any number of reasons to think of themselves as superior. Look at how they have been using eleven of Earth's languages, speaking and writing them as well as any well-educated native speaker. Also, if they have, indeed, studied every published document from the last hundreds of years, what does that say of their abilities? And their life spans—my goodness! From their description, they have a life span of more than ninety thousand years! Think of that."

President Kaitland said, "I'm taking all this in carefully. Can you please draw your conclusion for us?"

"Yes, thank you," Ellen replied. "I don't need to belabor it more. The point I want to make is that in the case of the Native Americans and the Europeans, all the many factors that they had in common meant next to nothing. The hierarchical culture and the personal ambitions of the Europeans overwhelmed any feelings of empathy, compassion, and fairness that they might have had. Now consider ourselves and the aliens. We don't share any common biology; and even if we did, we can assume that it would benefit us no better than it did the Native Americans with the Europeans. What we need to focus on are the cultural features and values of the aliens. What clues can we find in the things they have told us about their attitudes toward concepts like empathy, compassion, love, or fairness. I have been thinking about that and believe they have given us a great many clues. They've given us their story of overcoming war, of reducing their level of aggression, and of increasing their capacity for cooperation. They've spoken of lifelong loving pair bonds. I could go on with examples, but there is no need. I do tend to believe that an intelligent, rational mind would see the value of qualities like fairness and compassion—qualities we say we treasure. They have had much more time than we to refine and strengthen those qualities—time of both their culture and of their personal lives. Everything we have heard emphasizes how strongly they value those ethical qualities that we wish we possessed more completely.

Dr. Schneider asked quietly, "What if everything they have said is deception? What if all of it is designed to make us relax and drop our guard?"

"Yes, you are right to ask that—I was heading there next," Ellen answered. "I don't see how it is possible for us to discover the answer to that question without our being able to converse with them. If it is all deception and their motives are sinister and if they are as superior as they seem to be, then it is only a matter of time before they make their move. In that case there is little to nothing that we can do about it. On the other hand, if what they have said is true and they recognize and value us as another intelligent species, then developing a friendly relationship with them could be the singular most significant and positive event in human history. Oh, and one more thing. If it is true that they have watched us for 136 years and have judged that, despite our flaws, we have potential for good, I would love to be able to talk with them more about that."

President Kaitland glanced at Dr. Schneider again, feeling they had gained something worthwhile from this meeting. She said, "Dr.

HISTORY LESSON

Hazelstein, I have so enjoyed our conversation. Thank you for coming and sharing your insights with us. We will add all of these ideas into our consideration of the—"

Just then there was a knock on the door, and Eleanor Atwood came into the room. "Excuse me, Madam President. I'm sorry to interrupt, but the issue from the State Department demands your immediate attention. They are waiting for you in the Cabinet Room now."

President Kaitland responded, "Thank you, Eleanor. Please tell them I'll be right there." She turned back to Ellen and said, "Please forgive me; I must go. I still would like to have a little more time to talk with you. I recall that my staff arranged your flight to Tucson for tomorrow afternoon. I will think of a way to get in touch with you tomorrow morning before you leave. Thank you again. Goodbye for now." With that she stood and left the room.

Jonathan Schneider was also standing and reaching out to shake Ellen's hand. He said, "And I add my thanks too. I'm very glad the president thought to invite you here. You can count on her following through about getting in touch with you tomorrow. And now I must apologize and dash off to another meeting. I hope we will see you again here sometime. Goodbye."

Ellen was left alone with Eleanor who had returned to the room and said, "Nancy is available and can walk you back to the hotel or show you around Washington if you would like to see some sights."

"Thank you for such a kind offer," Ellen replied. "But I think I will wander around a bit on my own—perhaps a short visit to some of the Smithsonian Museums and then back to the hotel for a nap. I'm still feeling some jet lag. Could someone show me the way out? I'm a bit lost here."

"Yes, of course," Eleanor said. "Nancy is standing at the ready here and will escort you to the front lobby. Thank you again so much for coming today."

Nancy provided Ellen a few tips on things to see and told her to hang on to the coat—she would need it. As they said goodbye, Nancy handed Ellen another envelope with details of her flights to Tucson the next day. Ellen walked to the Smithsonian National Museum of American History a few blocks to the south and sat in the entry lobby to warm up. She looked at her watch and phoned Jim's number.

He answered right away and said, "I was hoping you would call soon! Gerry will be here in a little while to take me to the Canberra airport. How did it go, honey? Are you finished with your meeting already?"

"Yes, and I'm so relieved," she said. "It's all finished. Everyone including the president treated me so graciously, and they seemed to be interested in what I was saying. I think it went well, and now that I can relax, I'm just so tired."

"Yes, you must be; just take it easy now and rest. I knew they would be impressed with you," Jim said. "Do you know yet when you will be able to leave Washington?"

"They just handed me the details as I was leaving the White House. I haven't looked yet. Give me a sec." Ellen opened the envelope, read the itinerary and said, "This looks good. I leave Washington tomorrow in the early afternoon and need to change planes once, but I'm set to land in Tucson at 5:15 p.m.."

"Wow, that's amazing good luck!" Jim said. "I'm set to arrive in Tucson a little after 3:30 p.m. on my plane from LA. We'll be able to catch the same taxi together. I can hardly wait to see you and hear the whole story of everything that's happened."

They chatted a little longer until Ellen decided she needed to rest. They said their goodbyes, and Ellen added, "Say goodbye to Gerry for me and thank him for looking after my husband so well." Ellen felt elated as she pictured their reunion in Tucson the next day.

She managed just twenty minutes in the museum when she felt so tired that she gave it up and went back to the hotel. The prospect of an afternoon nap seemed wonderful.

Chapter 12

SEEKING CONSENSUS

Washington, DC, Tuesday, 30 January, 3:00 p.m.

At that moment, President Kaitland was still in her meeting with the Secretary of State, Devya Varma, and with the Secretary of Defense, Oscar Wainwright, plus a number of high-level staff from each of their departments. She had met with these people the previous afternoon to discuss General Beckworth's ideas and had asked Oscar Wainwright to talk with others from his department who had been at the morning meeting with Beckworth. She had already talked with Secretary Varma about the best way to approach their NATO allies and the UN over how to manage negotiations for an answer to the aliens' request. The current meeting was called when the Secretary of State was surprised to receive such quick responses to her initial feelers yesterday.

Secretary Varma said, "We've been in discussion with Britain, and the European Union looking for strategies to deal with the aliens. They are just as frustrated as we are of waiting for the aliens to do something. Well, the majority opinion is that the aliens have now done that something with their request. They have moved a piece on the board, and the Europeans want to get started with our next move. The people we have talked to so far take the same view we do—that the answer to the aliens will need to be issued by the United Nations. The aliens have not addressed their request to any one nation or group, and they have not given us guidelines in how to answer them. But it is our guess that they would only respond to what they see as a consensus from Earth."

"Have you received any response yet from the UN Secretary-General or from China, Russia, or India?" President Kaitland asked."

"Yes and no," said Varma. The UN has answered that they are contacting the Security Council to deliberate, India has said it is considering the question now, and Russia and China have not yet responded."

Kaitland turned to Oscar Wainwright and said, "Have you and your people come up with a plan about how we can share General Beckworth's defense strategy with critical partners and still keep it secret?"

"We're not there yet, Madam President," Wainwright answered. "So far we have outlined the plan in high confidence only to Britain and Germany. They both liked the plan but surprised us by taking the view that the plan does not need to be secret. They shared their reasoning with us, and I must say that we found it attractive. They said that if Earth decides to invite the aliens closer, we should inform the aliens that we can only invite them close if we have means to defend ourselves if necessary. We would emphasize that they are strangers to us, and we only have their word that they mean us no harm. They, in turn, would need to take our word that we mean them no harm while they come close with what amounts to a diplomatic overture. That's the way it is done by convention everywhere on Earth. Any diplomat or head of state that visits another country enters that country's sovereign territory and expects to be surrounded by that country's defenses—military personnel—guards around the palace or whatever. The aliens know so much about us that none of this should surprise or offend them."

"That's a very interesting point of view," the president said. "It would make it easier for us to approach Russia and China as potential partners for implementing the plan without fear that it might leak and become public knowledge. We could create a much more effective defense if we could partner with Russia and China. And some members of the UN might never be willing to consider inviting the aliens unless we actually do have a serious defensive capability."

The president paused and realized that she had been influenced by Hazelstein's words about inviting the aliens to come closer. She also sensed that the people in the room also favored issuing an invitation. She wasn't sure whether it was human curiosity to see what would happen next or whether it was frustration with an unknown period of passive waiting. She wondered whether the public at large would prefer to get on with it and see how the aliens would react to more direct questioning.

She said, "It's true the aliens acknowledged that Earth may need a significant period of time to decide the answer to their request, and that's a good thing because we will need all the time we can find to make preparations if we say yes. But I still want to move quickly on the process of placing the question before the UN for consideration and decision. Secretary Varma, please make that goal your department's highest priority. I expect that the UN General Assembly will drag out debate on this, and I want to get it started as soon as possible. It looks as if the Europeans will support us in moving this onto the UN agenda. Continue probing Russia and China—they will be critical."

"Secretary Wainwright," she said, "ask the Joint Chiefs of Staff to have their intelligence arms compile a list of Russian and Chinese hardware that we know of with the capabilities we need for General Beckworth's idea. I would like to have that information by this time tomorrow."

"Yes, Madam President," Wainwright answered. "That task is already underway and will be ready in time."

"Good." The president went on, "And we should think through the logic of the British and Germans of being open about the plan. I'm inclined to think they are right. We need to remind ourselves that the aliens are listening to all Earth communications that they can intercept. Chances are that we could not maintain secrecy for such a complex plan even with our best efforts. Also we can assume that they have detection technology that would alert them to the presence of the various missiles we would want to deploy in high Earth orbit. If we pretend that we have not developed this weapons system and they learn of it, they will know we have played falsely with them. That would not be a good start to a relationship to say the least."

Addressing both Wainwright and Varma, she said, "Would both of you please consult with your strategists and have them put together a pro and con list for me by 1300 tomorrow that explores this question. This issue of secrecy or not needs to be decided as soon as possible; we'll need every bit of time we have to put all the parts together." The president adjourned the meeting and went to another with her chief of staff to discuss which congressional committees they would need to brief first.

Washington, DC, Wednesday, 31 January

Ellen awoke at 3:00 a.m. the next morning surprised and pleased that she was able to sleep so long. She showered and dressed and decided to splurge and have room service bring her a carafe of coffee. When it was brought up, she learned that everything for her was already paid for, a happy surprise. As she sipped it gratefully, she thought about her meeting yesterday at the White House. She found it difficult to believe that President Kaitland did not already have very capable advisors who would have told her the same things that she did yesterday. Perhaps though it hadn't emerged yet from internal discussions or perhaps it helped to hear confirmation from an outside source. In any case, Ellen was relieved that both the president and Dr. Schneider had been receptive. She wondered what the president meant about getting in touch with her again and what she would like to ask her.

She put on the borrowed coat and went for a short, brisk walk around nearby blocks. The streets were still pleasant and peaceful before dawn. Then she returned to the hotel and packed while waiting for her breakfast to be brought up. The breakfast was good, and she was finished with it by 7:00. Just as she was beginning to wonder what to do next, there was a knock on her door. She opened it and saw Frederick and Sarah.

Sarah said, "Good morning. We're sorry to disturb you this early. We'll just need a moment of your time."

"That's quite all right," Ellen said. "Please come in."

They stepped in but only wanted to give her a brief message. Frederick said, "The schedule you received yesterday mentioned that we would pick you up to take you to the airport at 11:30 a.m. The President's Office has asked us to pick you up earlier at 10:15 a.m. and take you first to the White House for a short meeting with the president. After that meeting, we'll take you directly to the airport."

Ellen was surprised but saw that the president had followed through on her last comment yesterday. "Yes, I can be ready by then," Ellen said. "I'm already packed."

"Excellent," Frederick said. "We'll knock on your door again at 10:15. Goodbye until then." Ellen sat down wondering again what the president might want to discuss with her.

At 10:15, Ellen heard the knock. She opened the door, and Frederick, eyeing her closed suitcase approvingly, said, "I see you're ready to go." He picked it up and escorted Ellen to the waiting car. He took her to the lobby entrance of the West Wing and as he opened the door for her said, "I'll be waiting for you here when you finish. See you then."

Nancy Birchfield had already come out of the lobby to meet Ellen. Nancy thanked Frederick and escorted Ellen inside, through security and on to Eleanor Atwood's office.

Eleanor said, "Welcome back, Dr. Hazelstein. The president is just finishing a phone call and will be able to meet with you soon. Please have a seat. Would you like a glass of water?"

"Thank you, no," Ellen said. "I'm fine." She sat down with her mind racing.

After five minutes, the door to the Oval Office opened, and Rachel Kaitland waved to Ellen, "Please come in. Thank you for coming back this morning."

In the Oval Office President Kaitland gestured to a comfortable-looking chair for Ellen and then sat down in another facing it. Ellen noticed that there was no one else in the office. The president began, "I still have something I wanted to ask you, but that urgent meeting yesterday interrupted us. I agreed with you that it is the culture and attitudes of the aliens that we must focus on. And I agreed with you that if their motives are sinister, then refusing their request to come closer would not in the long run protect us. We still have no way to know anything with certainty. But I sensed from you that you are inclined to trust that they are being honest and that their motives are friendly. What I wanted to ask you is can you explain more fully why you feel they can be trusted? Can you identify particular points that have reassured you?"

Ellen was surprised and flattered that the president would ask her this. It would be only her opinion and not something from her field of expertise. She said, "Thank you for giving me the chance to express my feelings. This question has preoccupied me in the last several days. I tried to focus on things I could have confidence in being true—points not based on what they have told us about themselves. First, they are here. Even if they lied about how far they have come, we can still be confident they are at least from another star. And even the nearby stars are vastly distant and far beyond our present capabilities. From that I ascribe to them qualities of significant intelligence even if they did not develop the technology themselves but bought or stole it. Of course I'm assuming that operating and maintaining an interstellar spacecraft would require significant expertise. And yesterday I mentioned that their use of our languages also suggested a very capable intelligence. Yes, it's true that intelligent individuals can be psychotic and dangerous, but a culture shaped by such beings would not, I think, be stable over the long-term, and such technology would seem to require cooperation over a long period of time. Beyond this point, my logic softens into something more like hopeful speculation. Cultures of intelligent beings would see and understand the value of cooperation and I think would ultimately embody cultural values of reciprocity and fairness. A society just works better that way.

"Second, I ask myself what motives might the Visitors have for wanting to invade us or destroy us. Earth has no material resources that they could not find in many other uninhabited systems closer to them. The distinctive thing that makes Earth different is all the lifeforms here. It's implausible to me that they might want to farm us or enslave us for some bizarre purpose. And I think there's even less chance that they are some kind of sadistic culture that would take pleasure in watching us suffer or find pleasure in seeing us destroyed. Yes, I concede it's possible that they might see us as objects to study or even as objects of entertainment. It's also possible that they could see us as a potentially dangerous infection on the planet that should be destroyed before we become dangerous to them. I can't rule out the possibility that a successful culture of intelligent beings would think only of themselves as being valuable and think of other different beings as objects for which they cannot feel empathy. But if they had those kinds of attitudes, would they have the perspective or even the thought to talk as they so often have about empathy, compassion, and so forth? Yes, my analysis is the product of a human mind. There may be much more to consider that is beyond me. What I have just described along with the fact that they have made no move against us and have requested our approval rather than made a demand—these are the things that help reassure me. The way they have described themselves and their culture is so consistent with my idea of an advanced society. But they are intelligent, and it could be a clever ruse—they might be lying about everything. They might have bought or stolen their technology from a different civilization. Still,

I find myself drawn into their words. I'm guilty of wanting those words to be true."

"Thank you," President Kaitland said. "I see we both have thought about this issue in similar ways—ways we reach for when facts are few and far between." She paused and looked aside for a moment. Then she turned to Ellen and said, "Yes, that's all I need for now. If you will indulge me, I may want to get in touch with you again in the future. It's very good for me to know of sensible people I can call on in times of need. In the meantime, I know you have a plane to catch, and I have a very busy afternoon ahead." She stood and led Ellen to the door. She took Ellen's hand in both of hers, looked into her eyes and said, "I hope you have a pleasant trip back to Tucson and a very happy arrival home. Thank you again so much."

The president's kind parting words were flawlessly fulfilled. Ellen had an on-time, routine flight home and a happy and relief-filled reunion with Jim. Even though it was a time of day when many flights were arriving, the self-driving taxi rank had not yet emptied, and they were able to secure one right away. They chattered nonstop all the way home. Ellen was still describing details of her time in Washington as the taxi turned into the driveway of their modest home on the north edge of Tucson in the Catalina Foothills. It was built on one of the higher ridges and gave them wonderful views of the spectacular sunsets so common in Tucson. Just off their backyard, there was a large, deep draw filled with desert cacti including the giant saguaros as well as scattered mesquites and palo verde trees. It was a perfect home for numerous doves, quail, and cactus wrens. Groups of deer and squadrons of javelina sometimes wandered into their backyard and sometimes a bobcat too. Jim and Ellen were happy to call it home.

Washington, DC, Thursday, 1 February, 8:00 a.m.

President Kaitland was holding her regular morning meeting with her intelligence and defense analysts. She had also invited State

Department people to brief them on international developments of the last two days.

It happened that Tom Beckworth's friend, Joe Garcia, had been selected to brief the meeting on his technical group's report of suitable delivery vehicles that could be converted into high-orbit defense missiles. President Kaitland decided that Joe's report would be the best way to start the meeting and bring everyone up to date with what they might have to work with.

She opened with, "Thank you all for your prompt arrival. We have a lot to cover this morning. I would like to begin by calling on Space Force Colonel Joseph Garcia to brief us on our current inventory capable of delivering substantial payloads from high Earth orbit. Colonel Garcia, you have the floor."

"Thank you, Madam President," he said as he stood and walked to the front of the room. This was Joe's first direct briefing to a president, but he wasn't nervous. He knew his subject extremely well and had plenty of experience lecturing and giving talks before high-level audiences. He began, "In case any of you have not already received a copy of my group's written report, I've put extra copies on the end of the table. Fighting a defensive war in high Earth orbit regions would be something new for us. We have never built delivery platforms for this purpose. Therefore we must adapt existing vehicles that were designed for extreme high-orbit placement or for lunar excursion missions. We have also looked at Mars excursion vehicles but feel that they would be a second choice due to their small number and greater than needed size.

"It's fortunate that about twenty years ago, the US began developing orbital fueling stations in partnership with commercial companies. One of the critical technologies developed in that process was learning how to store large amounts of liquid fuel and oxidizer in orbital storage tanks for long periods of time. For these new defense missiles to work as needed, we can't use solid fuel rockets but need the flight flexibility of liquid fuel rockets. We will need to augment much of the existing fueling infrastructure so that we can place each delivery vehicle—I'll call them DVs—in its assigned position and then fill its tanks to leave it waiting and mission ready. It's doable because we can move much of what we need from other assignments. At this moment, we have identified only nine DVs that can be modified to fit our requirements. There are two size classes – four DVs of a smaller type with each capable of delivering three MIRV warheads of the current W87-1 or W88 class and five DVs of a larger size each capable of delivering seven MIRV warheads of the same two classes. It's possible that during the next three months, one more of the smaller DVs could be made ready for deployment in time. By the way, all of these DVs will be adapted second-stage designs and already have nozzle configurations optimized to operate in vacuum. They will be lifted into low Earth orbit first by reusable, first-stage, heavy lift vehicles."

Tom Beckworth already knew all of this from his previous Saturday night talk with Joe. He asked, "Colonel Garcia, have your people been able yet to calculate the optimal configuration of the orbital positions of the DVs for the most efficient attack posture?"

"No, not yet," Garcia replied. "I have two groups working independently on that issue now. The problem is complicated by not knowing where Earth might invite the aliens to park and where the aliens do park after they arrive. We have picked several probable positions and are working out configurations for each. In addition we are working on configurations for a larger number of DVs on the possibility that one or more international partners join the plan and add some of their resources. We believe the algorithms we have created will allow us to make last-minute refinements after the aliens have settled into position."

Another twenty minutes of discussion examined whether additional orbiting infrastructure was needed and whether it could be put in place in time. The time needed was a glaring obstacle, but reaching the goal might just be possible.

Next President Kaitland called on two senior Foreign Service Officers to explain the most recent international developments. The first officer, who dealt with issues from the European Union, explained that meetings in the EU were ongoing, but reports indicated that there was a growing accord in favor of inviting the aliens to come closer. The whole EU had now been made aware of Beckworth's idea which had been named the "Earth Defense Shield" plan and soon shortened to the EDS. The EDS idea provided a strong measure of comfort to skittish nations that might otherwise have resisted the idea of inviting the aliens in. Whether the EDS merited that level of confidence was another matter, but it was being grasped as a means of being able to exert some measure of human control over the situation. Britain had already decided to come on board so with this news from the EU, a block of consensus was forming.

The second Foreign Service Officer's ambit was the Russian Federation, and she also reported intelligence from China and India as well. India had now been informed in high confidence of the EDS and appeared to be considering the idea favorably. Following personal phone calls the day before from President Kaitland to President Kuznetsov of the Russian Federation and President Zhou of China, the US had success in setting up high-level meetings for discussion. All sides agreed the issue was urgent, and the meetings had been arranged for the next day, one in Moscow and one in Beijing. The US planned to reveal the full EDS concept at each meeting.

Developments were happening faster than expected. President Kaitland had a separate meeting next with the US UN Ambassador and senior staff. Rachel Kaitland could see that she would need to accelerate the process of reaching out to the UN even more than she first had thought.

Chapter 13

INVITATION

Tucson, Thursday, 15 February

Despite the fundamental change to humanity's view of their place in the universe, people around the world had to continue making a living. For Ellen and Jim, the first week back from Australia had been hectic. They had been able to arrange substitute instructors for the term's first day in the last week of January. But on their return, they wanted to give their students their full attention. Their own graduate students with specific dissertation projects already underway had done well. But as feared, some of the students in both of their post-graduate courses had been derailed by the monumentally distracting events in the last week of January. They were dedicated teachers and pushed hard to bring their students back into focus. Though it made an awkward start to the term, they would never regret having made the trip Down Under.

By the end of their first full week back, they had restored order for their classes and managed to pull their own focus back to their work and not think of the Visitors constantly. By the end of their second full week, their routine was back to normal. Still, they recognized there had been a seismic shift. Nothing was the same anymore.

The changes were reinforced minute-by-minute in the media. Government officials in every national capital were wrestling with the question of whether the aliens should be invited closer or should Earth reject their request. Some countries actively sought the opinion of its citizens; others tried to suppress it in favor of a powerful elite's view. But somehow, a general dominant opinion of the masses percolated out and began appearing in media outlets all around the world. It seemed to be some sort of "wisdom of the crowd" phenomenon materializing. Pundits of every persuasion were talking about it and advancing a whole gamut of often contradictory reasons for why it was happening. Whatever the right reason might have been, the dominant opinion of the populace did not escape the notice of leaders of Earth's most powerful nations. It appeared that most people were in favor of inviting the aliens to come closer, and people were referring to them more often as "the Visitors."

Washington, DC, Thursday, 15 February, 8:00 a.m.

President Kaitland called her usual morning meeting to order. She said, "The situation is developing rapidly. There is substantial pressure now from our allies to finalize our position so we can make a joint presentation to the UN General Assembly. The Secretary-General has already placed the aliens' request on the agenda for the next meeting in less than two weeks. Therefore, this morning I would like to poll the outstanding issues and have your latest view on when we can be ready."

The Space Force confirmed the US could have its nine DVs ready for deployment in two months rather than three and that current operational warheads had been identified and were being transported for installation into the DVs. Sufficient technical personnel had been moved to the two launch sites, and installation of the armed DVs onto the heavy-lift rockets was expected to be completed in time. The optimal orbital positions had been finalized, and the main effort now was in refining the targeting guidance computer software for the DVs. Challenges remained, but personnel were optimistic. Further refinements could also be carried out remotely to some extent even after the DVs were in their assigned orbital positions.

The State Department group advised that both Russia and China agreed the EDS plan was a good idea. But each had decided to carry out a version of the plan according to their own specifications. Moreover, each would maintain command over its own forces and would not submit to a joint defense command. The State Department reported that for one major issue, they had found a surprising unanimity among all countries consulted. All agreed that the EDS did not need to be kept secret, and the aliens should be informed that it existed and why it was needed.

President Kaitland was satisfied they were on track. After a brief discussion and to everyone's relief, she adjourned the meeting earlier than usual. Everyone left the room quickly, anxious to get on to the many other demands of their day. Rachel Kaitland went to her private study just off the Oval Office for a few minutes of quiet concentration. She knew too well that everything was being rushed on an emergency basis. The urgency invited mistakes and an end result that might not work as planned or even at all. She asked herself again, why such a hurry? Why not let the aliens wait for an answer for a year while Earth does what it needs to do deliberately and carefully? She considered what she heard as the most common reason—the aliens said they needed to leave soon and might grow impatient with such a delay—it might provoke them—no one knew how they might react.

But she knew of two other reasons. She felt them herself—simple overwhelming human curiosity and frustration over standing by and waiting. People wanted to know more; they wanted to be able to talk with beings who could do what the aliens were able to do. And to do that, many people could deny their fear and choose to believe what the aliens said about themselves. *So simple*, she thought. The building momentum to issue an invitation seemed as if it would soon be unstoppable, and Rachel Kaitland didn't know if she would want to stop it. Her curiosity and frustration were as great as anyone's.

The only organized resistance that emerged came from several smaller fundamentalist religious groups who saw the aliens as an affront to their beliefs. Although the major religious groups had taken an approach of accepting the reality of the aliens, some of the fundamentalists called the aliens a trick from the Devil to tempt them into unbelief. Others called them an intervention from God to test their faith. But these groups were small and were ignored except as an aside in news reports.

The next two weeks were hectic for everyone involved in the development of the Earth Defense Shield. All knew that the question of the invitation had been placed on the UN agenda, and those working on the EDS hoped that the delegates would become bogged down in disputes as was common with other issues. They would have been happy to see months of delay that they could use for vital work. They desperately wanted more time and hoped that the alien vessel would require at least several months to make the trip inward toward Earth.

Most of Earth's population had heard how the best Earth rockets would need more than four hundred years to reach the aliens' current position and did not imagine that the aliens could make the trip in less than six months to a year. Those who remembered the alien's description of their vessel knew it could be done faster, but they still assumed the aliens would choose a slower speed within the Solar System and guessed three to four months. People working on the EDS estimated that after an invitation was sent, they would have at least three more working months plus the nine days it would take for Earth's invitation to reach the aliens.

On the political front, President Kaitland was surprised at how soon agreements were reached with the Security Council and the Secretary-General of the UN. In the last week of February, the question of the aliens' request was put before the UN General Assembly. There were no calls from any nations for postponement or committee formation to further study the question. In a surprise to nearly everyone, only two-and-a-half days were needed for short speeches in favor of issuing the invitation. On the auspicious date of 29 February 2052, it being a leap year, the vote was made and approval confirmed to issue the invitation to the Visitors.

The UN invitation included instruction for the aliens to position their vessel in an equatorial Earth orbit at one hundred thousand kilometers above the surface. It also informed the aliens of the existence of the Earth Defense Shield and explained the logic the British and Germans had first suggested—namely that visiting ambassadors entering the sovereign territory of another country always accepted the necessary presence of the host country's defense forces and trusted that their safety would be honored. Within hours and with an unprecedented efficiency, the formal invitation identified as a collective one from all of Earth was beamed outward to the aliens. Everyone knew that Earth would hear nothing in return for eighteen days.

When news outlets announced the invitation had been sent, there was elation in some quarters and sudden dread in others. The reality of the invitation was more confronting than the idea of it had been. But there also were optimists who were convinced that the aliens were friendly and that if Earth could talk with them, humans would be able to persuade them to give Earth some of their advanced technology. There had been hints of that possibility in some of the alien informational broadcasts of the last few weeks. But for those with a gloomier cast of mind, they were forced to resign themselves to the now evident reality that the aliens would arrive on their doorstep. That group at least found some comfort in the soon-to-be operational Earth Defense Shield.

Even for the optimists, though, there was a lingering and sobering thought. Now at the end of February after four weeks of broadcasts of the aliens' video records, one fact had been made even more clear than before. The astounding videos of Earth and the other planets plus libraries full of detailed scientific data about all the planets and even areas in both the Kuiper Belt and the Oort Cloud where humans had never been—no doubt was left that Earth was facing a civilization of immense capability and power. No matter how Earth had answered the aliens' request, the aliens could do whatever they chose to do.

※

Soon already far from Earth, the invitation was racing outward—ripples churning across an invisible pond at three hundred thousand kilometers per second. Speeding past the orbits of Jupiter, Saturn Uranus and Neptune, it then left Pluto's odd orbit far behind—still just beginning its journey. And then far beyond the planets and past the rubble of the giant Kuiper Belt, it was still flying fresh on the first day of its mission. Far far ahead it would traverse an empty zone where distance seemed to have no meaning. Still farther yet lay a barren zone where the Sun is seen only as a bright star. It is a place of stark, cold beauty, and if anyone were there to notice—a place of unutterable solitude—an emptiness tempered only by the dazzling stars of the vast firmament all around.

But a longer look would reveal a large spherical object waiting in its own lonely orbit around the distant Sun. Human eyes would overlook it; its black surface seemed to reflect nothing at all as if it did not want to be noticed. A very sensitive instrument would be needed to reveal a few unnatural forms extending from the sphere's surface. In some of those forms, that instrument might sense a temperature higher than the near absolute zero of space as those forms guided the energy of the sphere's own ripples beaming inward toward the Sun.

And somewhere in between, Earth's outward-bound message pierced through the aliens' inward bound broadcast—each forging an independent path toward waiting listeners.

Tied by forces greater than themselves, the once confident masters of the small blue planet huddled near the campfire of their star—waiting … waiting for a future more mysterious and more uncertain than any they had ever expected.

Chapter 14

MORE WAITING

Tucson, early March

Earth had sent her invitation, and a new waiting game began. Everyone understood it would be eighteen days before they could have a response. It might only be a cessation of the current regular broadcasts, but everyone hoped for something more like when they would start their journey sunward and how long it would take to reach Earth. Most made a mental note or even marked on a calendar the date eighteen days hence even though they knew that on the few days before that date, they would hear about nothing else from news sources.

Curiously, something new was in the media and in general conversation in most languages. There were common references to "Earth" doing this or "Earth" deciding that and references to the people of Earth being called "Earthlings." Some of those who had already decided that the aliens were truly friendly believed that their announcement of their presence was all part of their plan to unify Earth as a step toward eliminating its wars. These people believed the shift in language to "Earthlings" was just the first sign of the alien's plan having an effect. It was a reassuring thought that was gaining notice.

The Visitors' regular broadcasts had been running unbroken since the day they first made their request to come closer four weeks earlier.

The amount of information sent so far had been astonishing in the quality of the graphic images and in the depth and richness of added scientific detail. News readers and writers had exhausted their vocabulary wearing out adjectives like "amazing, "stunning," "jaw-dropping," "staggering," "astounding," "dumbfounding," and so on. Scientists and historians studying the broadcasts agreed that even those adjectives did not do justice to what they were seeing. The Visitors, after saying that they would not share their advanced technology with Earth, nevertheless were showering humanity with detailed information about the Solar System that would have required hundreds of years or more for humans to gain on their own. Presumably the aliens thought such information was not dangerous and might encourage humans to do something more constructive than fighting with each other.

In Tucson, Jim had managed to get back into his research, and Ellen had gone back to writing a book she had begun the previous year. At the beginning of February, they had begun regular video sessions with Sandra and Gerry in Canberra. Weekends worked better since they needed to account for the big time difference. The friendship and closeness they had solidified in Canberra had not faded.

Needless to say, Sandra and Gerry were eager to hear what happened with Ellen's meeting with the president. Except for some of the president's specific responses, Ellen recounted all of it. None of what transpired was an issue of national security. Nothing Ellen had said would need to be classified, and Ellen was still surprised that the president had wanted to hear her opinion. In the end, she was grateful for the experience and was very impressed with President Kaitland.

They continued to discuss their fascination with the Visitors and the exciting new things that appeared every day in the broadcasts. The Visitors had begun sending practical information volumes about particular aspects of the Moon and Mars. Jim and Gerry thought they were enticing humans to do something with the Solar System by providing helpful hints. The Visitors had already mentioned how lucky humans were to have the Moon as such a valuable nearby base to help

MORE WAITING

launch themselves into space. For Gerry, this kind of information was like candy to a child.

He had been studying a recent broadcast that contained a large amount of data for the Moon, and he could not contain his excitement. "It's just amazing … it's fantastic!" he almost shouted. "I can't believe what they're handing out to us—it's incredibly valuable. Several days ago they sent a plan for a complete lunar-based electrical grid powered by solar panels that were arranged around the Moon's equator in an optimal way according to the specs. And it's achievable too—not some kind of crazy scheme that needs to cover immense areas; it's just a band around the Moon's equator. The design of the solar panels looks ingenious including modulator systems and some kind of clever-looking motorized gimbal mounts that will track the Sun. The specs said they're located and arranged for highest efficiency. Just think, a day on the moon is two of our weeks. At sunrise for any panel, it perks up and aims at the Sun and tracks it face-on for the next two weeks until that panel passes into a two-week night. But never mind; other panels are in full sunlight and others are emerging into their mornings. Since half of the Moon is always in full sunlight and since the panels are all around its circumference, the connected grid will have its full nominal power every minute unless you want to modulate it downward for some reason. That would be easy to do just by shifting the aim of the panels. There's no atmosphere and no clouds to reduce the sunlight—just clean vacuum all the way up. And in the Moon's low gravity, all of those structures would be easy to build, and they'll never need to be reinforced against winds. And wait, get this—I've been in touch with a friend in Melbourne who knows a lot about solar panels. He said the specifications described a design change to the chemistry of the panels that stated an improved efficiency of better than 92 percent over our current best designs. That's huge—enormous!" Gerry's excitement was peaking. "My friend worked through the detail and said the total power output of the Moon system as designed would surpass the total present electric energy output of all of Earth by 370 percent!" In Gerry's video window, he was waving his arms wildly. In the other video window, Sandra was laughing.

Jim and Ellen were laughing too, but with pleasure at Gerry's exuberance. Jim was just as excited about the design but didn't get quite as worked up as Gerry. It would be an amazing resource, but that would be a long way off. "Gerry, I'm with you, man," Jim said. "It's fantastic! We need to see about getting jobs on the Moon."

On the next call, Gerry opened with, "Guys, have you seen what started broadcasting night before last? I've had a chance to look through it now, and oh my gosh—it blew me away!" He was already starting to wave his arms. "There were two main parts. The first was a map of the richest metal deposits on the moon down to a depth of about four hundred meters beneath the surface of the major lowland maria. You knew already that the lunar rocks brought back to Earth eighty years ago had assayed with good amounts of aluminum, iron, magnesium, and titanium, right?"

Sandra broke in, "Sure, I knew that!" Everyone laughed.

Gerry, laughing too, went on, "Well they found lots more! They're all oxides sure, but that's no problem with all that free solar energy available. After processing the ore, you wind up with enormous amounts of refined metals plus enormous amounts of pure oxygen to be used in habitats and ... well ... lots of things."

"Calm down, Gerry," Sandra laughed. "I don't want to have to fix you after you tear something loose inside."

But he was unstoppable. "And the second thing they included is big ... really big! You guys know about rail launchers, right?"

"Uh oh, we're off again," Jim said laughing.

They all laughed. Gerry always put them into a happy mood.

Jim said, "OK, Gerry—how big?" And he thought—*How I wish we were sitting around a kitchen table again. It was so much fun.*

"OK, OK. I'll try to calm down," Gerry answered. "But I mean big—*huge*! So back to rail launchers or mass drivers as they're sometimes called. They're basically linear electric motors; the armature runs along a double rail instead of spinning in place. It's an electric catapult that can throw projectiles much faster than an exploding chemical propellant in a gun barrel can. They do need a lot of electric power, but if you have that, you can scale them as big as you want. Now picture one of these on the Moon and scaled to be two or three kilometers in length. There's no atmospheric friction to worry about when you want to launch a package at very high speed, and I'm talking *fast*—like fast enough to escape the Moon's and Earth's gravity and able to travel all the way to an orbit around Mars. And you can vary the launch speed to choose different destination orbits by varying the electrical power. The aim could be optimized for standard destination orbits by having several rail launchers with different orientations and further optimized by choosing launch times when the Moon's rotational position is just right. They could be built big enough to launch a hundred thousand tons of payload in one shot. Multiple launcher systems could send millions and millions of tons of refined metals to any orbit wanted around the Moon, Earth, Mars, or the Sun. Probably farther out too if needed.

"The Moon could host an enormous metallurgy industry with alloy foundries and plate forming and beam forming factories. Just think, preforming facilities there could build structural components in whatever shape would hold up to the g-force that the rail launcher would exert when it shot them off to some destination. They could be sent into space with the push of a button and be welded into place on some huge project in orbit anywhere you liked. Think how much better it would be to get all that material in orbit wherever you wanted it for space stations and research labs. No need to haul it up with such difficulty and cost out of Earth's deep gravity well on huge expensive rockets."

Jim said, "I remember reading the military had been researching rail launchers for a long time, but I hadn't heard of any applied projects coming out except the ones from the Navy decades ago to launch airplanes off aircraft carriers. So what did the Visitors send that had to do with rail launchers?"

"Oh my gosh, I forgot to say the main thing!" Gerry laughed. "The Visitors included a detailed plan for a rail-launch system on the Moon for doing what I just described. It had all the technical designs, specifications, materials to use, information on optimal alloys for the structures—heck, I can't remember it all. Oh wait, another big thing—no, it's huge too! They detailed an improvement to our existing designs that according to their specs would increase power and electrical efficiency by three-fold. It's mind boggling!" Gerry was starting to wind down, but his work was done. Everyone was very impressed.

Washington, DC, Thursday, 14 March

Two weeks passed in relative quiet after the invitation was sent. News outlets exhausted themselves with every speculation that fertile minds could create about what might come next. From outward appearances things seemed calm, but there were pockets of furious activity behind the scenes. Every government branch involved in creating the Earth Defense Shield was working at full capacity, not only in the US but in Russia, China, and Europe too. No one was sure just how much time they had. The job ahead of them was enormous and complex, and they had been given ridiculously short deadlines. Unexpected delays and breakdowns could make hash of their schedule.

When President Kaitland convened her 0800 meeting on March 14th, she was hoping to hear some good news. Progress had been going to schedule, but some supply problems for fuel delivery to the orbiting tankers had appeared and threatened one of their completion milestones. With only four to five days before a response from the aliens might arrive, everyone was feeling an increased tension. She no longer held these meetings on a daily basis. The strategic questions had been replaced by operational issues, and the president knew better than to micromanage those. She trusted the professionals in charge. Now she wanted to be kept up-to-date on progress and know that different operational centers were performing their respective roles according to plan.

General Beckworth still came to these meetings even though strategic problems were not in the forefront. With his advisory role at the Pentagon, he continued to have significant influence in a number of important areas at high levels. In particular he had been keeping up privately with his friend Joe Garcia and knew great detail about the hardware of the Earth Defense Shield. He was elated that his EDS idea had been taken up so widely. It exceeded all of his hopes, and he now felt as if humans had a chance to defend themselves if the aliens proved to be hostile. He was truly grateful to his friend Joe for pulling him out of his earlier lapse. Despite all the pressures still present, he now felt immense satisfaction over his plan and the fact that working to implement it was now his primary duty.

In the first forty-five minutes of the meeting, President Kaitland worked her way through most of the issues and heard good news. Every area was still on schedule, and the fuel supply problems had been solved. A lower level officer from Joe Garcia's group, one closer to operations, reported that the modifications needed for mounting the MIRV platform assemblies in the DVs was even a bit ahead of schedule. And crucially, the DV guidance systems were testing well to nominal specifications. President Kaitland felt like smiling at this abundance of good news but caught herself in time when she noticed General Beckworth indicating he had a question. She hoped it was not bad news. She said, "Yes, General Beckworth, you have the floor."

"Thank you, Madam President," Beckworth said. "Among all the critical issues we are dealing with, there is one I have not heard addressed yet. It is the issue of the separate command structures of the EDS for the Russian and Chinese missiles. Europe has agreed to merge their DVs and command with the US in positioning the coordinated orbital stations. Russia and China rejected that option, but I have heard nothing of how they intend to complement the US and European force to form an optimal defense posture. I fear that if this issue is not being addressed, there will be serious downstream problems. Is some other group dealing with this?"

There it is. President Kaitland was well aware of this problem and had been quite worried about it. The State Department had had no success in coaxing a commitment for cooperation from Russia and China. She wished he had not brought it up today, but she knew it was overdue. "Thank you, General Beckworth," she said. "That is an important and timely question. We had hopes for a breakthrough in discussions with the Russians and Chinese during this past week. But it hasn't happened. You're right that it should be settled sooner rather than later. With only one month left before deployment of our last DVs, time is urgent. We still do not know the number and type of DVs the Russians and Chinese will be adding to the EDS. I believe the orbital planning group has already developed a number of different placement scenarios anticipating this problem. Do you expect that it would require much more time to fit that unknown number of DVs into the plan?"

"I think no," General Beckworth replied. "Our group is comfortable with the plans they have already developed and can handle whatever the Russians or Chinese are likely to add. It's a different issue I'm worried about—namely how difficult it might be for us to realize when is the right instant that we must give the command to defend ourselves—what would trigger us to command a strike on the alien vessel? Please do not misunderstand me. I am not in any way referring to a first strike coming from us directed at them. I only want to feel confident that we will not be the victim of a first strike from the aliens. Since we know nothing about what they may have, we are at a disadvantage even as we watch them at close range. If we are able to recognize that they are beginning an attack, that's one thing. But if the beginning steps of their attack are somehow unclear—if we struggle in indecision for even a few seconds, it could mean disaster. Now add to that problem the fact that we will have three different command entities facing them. One might mistakenly decide to attack without adequate justification while the others hold back. Alternatively, one might attack justifiably while the other two remain indecisive. My point is that if we cannot reach agreement with the Russians and Chinese to join into one command structure, then we must at least reach agreement with them over what criteria we will use and at what trigger points we will decide to give a command to attack. All of us need to be operating under the same rules of engagement. For the EDS to be effective, we need to act in unison."

President Kaitland had worried about this too, and it was the main driver for the negotiations to persuade the Russians and Chinese to join the US and the Europeans. It was no surprise that the US and Europeans did not want to submit to a Russian or Chinese command, and the Russians and Chinese had the same view from their side. Negotiations were still underway.

President Kaitland had been wondering whether she should phone the presidents of Russia and China again herself. But she worried that extra personal pressure from her might backfire. "Thank you, General Beckworth," she said. "As always, you raise an important issue. I agree that we need to develop those rules of engagement as a priority. Can you begin the process yourself and identify a person or group most qualified to help you? And can you give me an idea of how long you might need to prepare a preliminary written plan? We will want to propose a refined plan as soon as possible to the Russians and Chinese."

"Thank you, Madam President," he said pleased at her response. "Yes, I will take on all of that and aim to have a time estimate and first ideas for you by this meeting time the day after tomorrow."

After the meeting was adjourned, Tom Beckworth left feeling exhilarated to be able to sink his teeth into something so important. And he couldn't help but notice how his feelings toward President Kaitland had changed. In January she had seemed imposing and remote as the president and his commander in chief. He had been impressed with her intelligence and competence. But after so much interaction since then in such stressful circumstances, he was seeing her as a person—and as a woman too, a very attractive one. That realization surprised him almost as much as it challenged him. Although he tried to put that aspect of her out of his mind, he found himself wanting very much to please her.

Two days later, General Beckworth was as good as his word. He presented the president with a written report that included preliminary decision criteria for further refinement plus the identities of members of a working group he had organized to focus on this project. President Kaitland thanked him sincerely. It was good news for her to have that project in competent hands, and she found herself looking forward to his comments and increasingly grateful for his contributions.

◈

The Visitors' broadcasts continued and still brought new treasures of information each twenty-four hours. One day they included data from probes that plunged deep into the atmospheres of both Jupiter and Saturn. While the probes descended, they sent back reports of pressure, temperature, chemical composition, electrical and magnetic field strength, and so on. The probes functioned much deeper into those immense atmospheres than experts on Earth thought possible, but eventually even the aliens' probes went dead. As to the moons of the two giant planets, that data had not arrived yet, but there was a teaser in the text lines that it would be coming. The Visitors included a close-up sightseeing tour over the surface of Pluto—the spectacular images appearing so real that one seemed to be standing on Pluto's surface. Planetary scientists on Earth could not contain their joy over the scientific detail included with the images.

Sometimes the Visitors returned to historic videos of Earth. They must have maintained their closer presence at least through World

War II and the early days of human's moves into space. The documentation during the 1920s through the 1960s was wonderful. One thing was evident however. Perhaps because the Visitors were so dismayed by Earth's propensity for war or because they felt they needed to document it, extensive videos were included of World War II scenes in Europe and Asia of burning cities and cratered landscapes. They even somehow managed to capture video of the actual blast of the atomic bomb dropped on Nagasaki, Japan, and the horrific aftermath. However, there was balance too with included scenes of the rebuilt and thriving cities they had earlier shown being destroyed. There were videos of three of the early Moon landings—July and November of 1969 and January of 1971. They must have had no difficulty noticing the traffic of American and Russian space probes traveling to the Moon during those years.

After that third Moon landing, no more close-in videos of Earth or the Moon appeared. Apparently that was the time the Visitors decided they needed to move farther away from Earth. Incidentally, the Visitors' images of the Moon landings put to rest the claims still persisting that the landings were all faked. Nonetheless, a few resisted all evidence and insisted all of the Visitors' broadcasts were faked too. After the last of the Earth and Moon-focused broadcasts, the Visitors continued to send what seemed an inexhaustible supply of Solar System data that as always defied the adequacy of adjectives in any language to praise enough.

On the few days before the end of the waiting period for the aliens' response, news outlets everywhere were running red hot with anticipation. Various polls taken since the Visitors first made their request showed a substantial majority in favor of their approach. But there was still a significant minority who feared what might come next. And there were plenty of media personalities who were experts at making money by fueling and capitalizing on fear. Many heated and irresponsible claims characterized those few days. Tension was rapidly building.

International Date Line, Tuesday, 19 March, near midnight

As March nineteenth shifted to March twentieth at midnight over the International Date Line, the alien informational broadcast abruptly ceased, and a new screen appeared with the words, "Stand By Please." A half-hour later, a new screen appeared, and the familiar gentle voice began to speak the words scrolling on the screen.

> "Dear People of Earth, we thank you for your kind invitation. We are now preparing to begin our journey toward you. We will cease our regular broadcasts during the journey, but when we are closer we will again reach out to you. We look forward to many pleasant conversations with you after our arrival."

The screen text and the voice message began to repeat and continued for twelve hours when it ceased. The screens tuned to the Visitors' frequencies went blank for the first time since the broadcast of the first message was interrupted for two hours on 23 January before resuming with the Visitors' first answer.

◈

Nine days before Earth's people could listen to the Visitors' response to the invitation and far far away in the empty zone where the Visitors' ship followed its solitary orbit, the invisible electromagnetic ripples beaming out from the black sphere ceased. In the quiet of deep cold vacuum there was no stirring except for one visible change. The sphere began to brighten. Its darkness faded away and lightened until even in the feeble light of the distant Sun it floated bright in a silvery metallic sheen. It was as if it had preened for a proper and impressive introduction. And then it began moving silently and gracefully away from its orbital path. There were no violent eruptions of rocket exhaust, no vibrations, not even light—only the eerie silence of deep space. But if the right instrument had been there to observe, it would have registered great disturbances in a hidden and mysterious field—disturbances that signaled the marshaling of immense energies within the vessel. The Visitors were on their way.

Chapter 15

JOURNEY

Washington, DC, Tuesday, 19 March, 9:00 a.m.

The Visitors' announcement followed by the break in transmission electrified Earth's inhabitants. They had grown accustomed to the steadiness of the Visitors' broadcasts, and now even the optimists felt unnerved. Issuing the invitation nineteen days earlier was jarring enough, but now that the alien vessel was on its way to Earth, stark reality set in. The imaginations of those still suspicious of the aliens were unleashed with some people left floundering in their fear of the unknown. A variant of "buyer's remorse" began to appear in the media with commentators asking whether Earth had made a mistake after all by inviting the aliens closer. Second thoughts dominated the news for the next full week.

The biggest issue for governments was that the Visitors had provided no estimate of how long their journey to Earth would take. When the UN so quickly made its decision, it had thrown EDS planners into disarray. Everyone involved in the Earth Defense Shield worried that their estimates may have been wrong and that the Visitors might arrive sooner than they had planned. Everyone hoped the Visitors would resume sending messages soon. They knew that the Visitors could have begun their journey nine days before their response was received on Earth. How fast they were traveling and how far away they might be now though was unknown.

At her morning meeting just after the Visitors' response was received, President Kaitland asked how many of the Defense Shield DVs would be at their assigned orbital stations and fully fueled in two weeks. The answer was a little better than she expected. Four of the smaller class and five of the larger class DVs had already been launched to orbit and would be moved into their final position, fueled and be operational in two weeks. The Europeans reported that they would have three DVs equivalent to the smaller class in place by then as well. The Russians and Chinese had not yet revealed their status but promised information soon.

Hearing this news, President Kaitland said, "I'm glad to hear that we're on schedule and will have at least that many in place by then. I've had an uneasy feeling that we might be relying too much on our original estimates of how long the aliens need for the journey. Is there any scope at all that even one or two more DVs could be ready in two weeks?"

Colonel Joe Garcia was present at this meeting and answered, "We've been worried about that too and have explored every possibility. But all operations groups are already working twenty-four seven and no additional trained technical staff or facilities exist. We have projected beyond that two-week mark though and are pretty confident that we can have one more of the small class of DVs in orbit and ready eight to nine days after two weeks from now. That would be 10 or 11 April when we would have a total of five of the smaller class and five of the larger class in place plus the three European DVs."

"Lucky thirteen," President Kaitland said. "Thank you, Colonel Garcia. I'm glad to know we'll have at least that many arrows in our quiver. Let's hope it will be enough if we're surprised. And let's hope the Russians and the Chinese come through and join the party. The Visitors said they would reach out to us when they were closer. I hope they will alert us to when they expect to arrive."

Progress on the EDS continued on schedule with every operations group working at full capacity. It was unheard of—even ridiculous some thought—to plan, construct and deploy a crucial weapons system of such complexity in such a short amount of time. More than a few major players in the enterprise were skeptical that it could be put in place without serious accidents due to haste or without paralyzing errors that could cause the system to fail. There had been no time for a normal test program, and there was more than enough worry to go around. But there was no alternative except to try.

Washington, DC, Tuesday, 2 April, 9:27 a.m.

The silence was broken at last two weeks after the Visitors' response arrived. The primary radio frequencies the Visitors had been using sprang to life at an unusual time for them—9:27 a.m. in Washington. A simple text message scrolled on screens in all eleven languages and repeated for six hours after which the signal ceased.

> "Hello People of Earth. We are now approximately halfway in our journey to you and expect to arrive on 15 April. We will assume the orbit you specified. When we are closer and have slowed our speed, it will be easier to communicate more. We will reach out again at that time."

Suddenly everything changed.

The reaction to the brief message was a testament to how completely the Visitors preoccupied the consciousness of Earth's population. With Earth's people having gone through shock after shock during the last three months, this latest short message still had the power to galvanize them to new heights of emotion—everything ranging from profound dread to euphoric anticipation. One might have expected some habituation to have taken hold by then, but evidently not. There were some pockets of panic, and in a few places mobs rioted and damaged government buildings. The crowds surging into city streets weren't as large as in the first days of January, but they were vocal in their demands for government guarantees that the Earth Defense Shield would fully protect them. Once again religious leaders and government officials called for calm. And those government officials who were directly involved in deployment of the Earth Defense Shield felt the greatest pressure of all.

President Kaitland called an emergency meeting of her newly established EDS Operations Oversight Group as soon as the latest message was seen. Within an hour, all who could come were at the meeting. President Kaitland opened with, "Well, we have certainly been called to attention. Now we know how much time we have left, and it's the surprise early arrival we feared. Thankfully they'll arrive after our estimate of April 10 or 11 for thirteen defense missiles to be in place. Colonel Garcia, does that estimate still hold?"

"Yes, Madam President," Garcia answered. "We are even more confident now that we can make it. There is also a new possibility that the Europeans can add one more of theirs in position by April 13 or 14 to raise the total count to 14."

"That's excellent news, Colonel! Let's hope the Europeans have good luck too." She continued, "In the last week, we have received confirmation that the Russian Federation and China will each join the EDS but not under our command structure with the Europeans. Each decided to maintain their independence, and each will operate under their own national command. That means we will have three command structures involved. So far they have not specified how many units each can contribute, only a range. Russia projects nine to fourteen, and China projects five to eight. We will press them and hope for a precise number soon. The alien message this morning ought to give them a strong push." She turned to Tom Beckworth. "General Beckworth, what is the latest from your development team for the command structure rules of engagement?"

Beckworth responded. "We developed our draft version of the decision criteria and have discussed them with the Russians and Chinese. We expect them to send us their suggested changes by tomorrow. The military people and defense representatives there that we've been dealing with have been cooperative and seem sensible. It might need one more round of revision, but I'm optimistic that we can have an agreed version in less than a week. Then it will need to be approved by each nation, but the deadline we now have should make that a rapid formality. After it's finalized, we will need somehow to quickly embed those rules in the minds of everyone involved in the command structure of each nation. We have at best thirteen days before they are in orbit over our heads. Time truly will be short for that last step."

"Thank you, General Beckworth. That brings me up to date," President Kaitland said with relief. "All of us hope this system will never be needed and, God forbid, not at the first moment the Aliens arrive here. But we cannot take a chance and must plan for the worst. Please advise me at once of any serious issues. Thank you all."

Canberra, Wednesday, 3 April, 7:10 a.m.

When the aliens' midflight message came in, Gerry and Sandra were already asleep for the night. But the next morning, their phone sounded early. Gerry had just poured himself his first cup of coffee when he answered to hear Jim's voice.

"Gerry … Gerry, have you heard?" Jim blurted. "Oh, sorry; hope I didn't wake you. Good morning—have you heard … have you heard the report?"

"Have I heard what?" he asked. "What's happening? Something about the aliens?" Gerry was instantly excited by Jim's tone of voice.

"Yeah, right—absolutely!" Jim answered. "It came in this morning our time. Listen—they said they were already halfway here—they said they would arrive here on 15 April! Can you believe it?"

"What?" Gerry shouted. "Oh my God—that's crazy fast! Did anyone say they picked up the direction of the signal source?"

"No, nothing about that," Jim said. "Just endless talk about when they will be here."

"Wow!" Gerry was on fire. "Oh my gosh! Thanks, Jim! I'm turning on the news now and will make a couple of phone calls. I'll call you back as soon as I get some answers. Man, this is something!"

"Thanks, buddy. I can't wait to hear what you find out. Talk soon!"

Gerry called his favorite go-to guy at the Canberra Deep Space Communication Complex just southwest of Canberra by the Tidbinbilla Nature Reserve. He hoped that his good friend Simon there managed to get a fix on the signal's direction. He knew Simon would likely have done it. Simon was a long-time fixture in the control room of the complex and something of an eccentric old bachelor who was accorded unusual privileges. He had set up a sleeping cot in a side storage room and had been living there for some weeks. Even though others were monitoring the control room twenty-four hours a day, Simon had set up an alarm rigged to wake him if a new signal came in. And yes—he had pegged it! Gerry was ecstatic.

After Gerry finished his call to the Tidbinbilla Space Complex, he put a cup of coffee in Sandra's hand, and soon they were on a video connection to Tucson. Jim and Ellen answered right away.

Jim asked, "What did you find out? Anything good?"

"I'll say! Oh, hi Ellen." But Gerry couldn't wait for any more courtesies of hello-type small talk. "Get this—my pal down at the big radio dish told me they were able to zero in on the radio source, and it's not where it used to be. Now it's shifted to 7.4 degrees off the ecliptic toward the Southern Celestial Hemisphere. Man, do you know what that means?"

"Uh, uh…" Jim stammered. "Oh! Yes! Right, now I see what you—"

"Yeah! You got it, Jim!" Gerry blurted without letting him finish. "That has to be it."

Ellen and Sandra were looking at each other as if to say—"*now what are they raving about?*"

Sandra broke in with, "What do you mean by degrees off the ecliptic? What's that?"

"Oh," Gerry answered, "the ecliptic is just a name for the planetary plane—all of the planetary orbits around the Sun lie pretty close into the same plane." He raced on, "Yeah, they obviously are not coming straight in from where they were. They've pulled out of the planetary plane toward the Southern Celestial Hemisphere. I'm guessing that's so they can skirt around the bulk of the Kuiper Belt and move into cleaner space. That way they could rev up their drive to make better time. They probably started out slow and will slow down again when they get close to us. But in between, they can really race. Just now I played with my calculator—I'd guess that in the main part of the run, they're cruising somewhere around a third of the speed of light. It's nowhere near the top cruise speed they mentioned before, but by our standards, it's an insane speed!"

Finally Gerry paused for breath, and they were all able to say hello.

But he wasn't done. "Do you realize what that means about the message they sent? If they're traveling that fast, the radio signal they sent would have been blue-shifted a lot for us here, enough that we might not spot it. But we received it on the regular frequencies we were monitoring. Their velocity vector would have been paralleling the planetary plane, and since they're well outside of the plane, their radio signal would have had to reach here at an angle from their direction of travel. So they had to calculate all that in advance and adjust for the Doppler shift just the right amount toward longer wavelengths so that we would receive it where we expected to find it. Well sure, these guys know what they're doing."

The rest of them started to respond when Gerry interrupted again. "I can't wait to see what their ship looks like. They were told to orbit a hundred thousand kilometers from Earth's surface. So that's about sixty four thousand kilometers above the high orbits of the geostationary satellites. I'll bet every telescope on the facing half of Earth will be looking at them when they settle into their parking place. I wonder if we'll be able to see their ship with the naked eye at that distance. Man, I hope we'll get some decent photos to study. This is all just too much—too much!" Even though Gerry just got out of bed, the rest of them thought he would need to lie down for a while to recover from the excitement.

Gerry wasn't the only one to work out how the Visitors planned their trip to Earth. The Visitors said the message was sent at the midpoint of the journey. From that clue, the 7.4 degree angle from the planetary plane told astronomers that the alien ship had moved out to about fifteen billion kilometers—one hundred astronomical units south of the plane. The main radio astronomy facilities in the Southern Hemisphere worked out an estimated range of directions where they would look for the approaching alien vessel's signal when it was closer to the outer planets. No one knew just where the aliens might decide to begin their arc back toward the planetary plane, but at least astronomers could make reasonable guesses and were confident someone would pick up the signal quickly.

During the next week there was frantic activity at every operations center involved with the EDS. President Kaitland continued her regular meetings with the EDS Operations Oversight Group and so far had not heard any significant bad news. Everything was still on schedule as projected the week before. General Beckworth had come through with the final version of the rules of engagement for the three command entities, and they were now in the hands of each of the national governments for review and approval. By the eighth of April, one week before the Visitors would arrive, the Russians confirmed that they would have nine, and the Chinese confirmed they would have five of their armed DVs fueled and in position by two days before the aliens' arrival.

The two words, "The Arrival," had become capitalized as standard usage everywhere in the media. Once again commentators were going hoarse with their discussions of what might happen and how things might change for Earth. Tension and anxiety were building after starting from an already higher than normal level. This would be like no visit in the past from some powerful head of state. No one grew tired of repeating that it would be like nothing else ever before in all of human history.

Earth's radio telescope installations in the Southern Hemisphere were scanning the expected range of directions to catch any new messages that might come in. Southern Hemisphere optical astronomy facilities like the Vera C. Rubin Large Synoptic Survey Telescope (LSST) were waiting for the first hint from the radio astronomers of where they should aim their instruments. The relevant space telescopes like the new Copernicus, the successor to the Webb, were also being made ready. Powerful radar installations that had been able to track distant asteroids were primed to reach out for the alien vessel when it finally came close enough.

Washington DC, Friday, 12 April

The days before "The Arrival" paced by slowly—the sky remained silent. The tension for many people was almost too much to bear. Then on Friday at 5:48 p.m. in Washington, DC, the silence was broken. The screens tuned to the Visitors' frequencies came to life with another scrolling text message and this time accompanied by the gentle voice.

> "Hello good people of Earth. We are now drawing close and have reduced our speed. We are still on schedule to arrive at our assigned orbital position on 15 April. Soon you will be able to see us. At the end of this message, we will attach specific coordinates for your Southern Celestial Hemisphere map so you will know where to aim your instruments to find us. When we are closer, feel free to reach out with your radar facilities so you can track us more precisely and feel reassured

that we are approaching safely. We look forward to our arrival and to conversing with you much more. We will sign off for now and will give you precise details of our approach during its final hours."

The voice went silent and below the text of the message were the promised coordinates in the Southern Celestial Hemisphere. It was a short message and repeated seventeen more times before breaking off and leaving the screens blank.

Within minutes at every significant astronomical facility in the Southern Hemisphere, operators were scrambling to reset their instruments to the specified coordinates. NASA's Mission Control Center was doing the same for the Copernicus Space Telescope which, like the Webb Space Telescope, had been stationed at Earth's L2 Lagrange point. The radio astronomers including Gerry's pal Simon had been monitoring continuously and were on it straight away. They were able to pinpoint the signal and confirmed that it had now shifted from the midflight angle of 7.4 degrees off the planetary plane to 91.9 degrees from the plane. Simon phoned Gerry right away. They were both puzzled by the new angle. They thought it should have been around seventy to eighty degrees because they expected the aliens to arc back toward the plane when they drew closer to the outer planets.

Then Simon said, "Wait, what if they stayed out from the plane at the hundred AU mark until they drew up even with the Sun and then turned sharply inward toward the Sun for a time before angling back a small amount toward Earth. That course could give us this new angle for the signal."

"But it's out of their way. Why would they go past Earth to be abreast of the Sun and then angle back toward Earth?" Gerry asked.

"Dunno," Simon said. "Maybe since they were way out there on the trip anyway, they decided to do some kind of broad survey of the Solar System. From one hundred AUs out to the side of the plane and opposite the Solar System's center, they could have gotten a good view of everything on all sides of the Sun out to two or three times the orbit of Pluto. And anyway, compared to their whole trip, Earth is so close to the Sun that they might have considered the overshoot a trivial excursion."

"Hey, that's a great idea, Simon," Gerry said. "With their technology, they might have been able to track the orbits of every kind of knickknack out there—like for instance things that might collide with Earth someday. I hope we'll find out what's going on."

They agreed that the aliens seemed to have no worries about fuel or efficiency when traveling in the scale of the Solar System—just head out there and go for whatever looks interesting with no mucking around. They both were dazzled by thoughts of what it must be like to be on board that vessel.

The recently upgraded Southern Hemisphere Planetary Radar System in Australia also had its antenna dishes aimed to the coordinates even though staff there knew the alien vessel would still be too far away for them to detect. Nonetheless, they sent a probe signal out at their maximum power level as a systems check and as a good baseline level test to be ready for the real event. To their great surprise, they received an echo a little more than ten hours later that was strong enough that it didn't need heavy signal-to-noise processing. The observatory staff didn't know what to make of it and worried about equipment malfunction. They knew of no object in that direction that could be reflecting a radar signal. If it truly was an object and not an equipment glitch, it was at a significantly greater distance than Neptune's orbit, the farthest planet from the Sun. Even more startling—the instruments indicated its velocity toward Earth was a tenth the speed of light—thirty thousand kilometers per second!

The news of this radar echo raced through the local network of observers. The news reached Canberra Saturday around seven in the evening. Gerry had already gone to the Tidbinbilla complex several hours earlier to study the previous radio signal data with Simon. When this news came in over their com link, they dropped everything and studied the message.

Gerry said, "How could they be detecting the aliens' ship at that range? Their systems were upgraded two years ago, but they can't be that good. It must be an equipment glitch."

"I just had an idea," Simon said. "The aliens seem so helpful; maybe they took our radar pulse, amplified it, and shot it back to us so the signal would be strong enough for us to see it. You know, a similar idea to an airplane's transponder. And maybe they got their Doppler shift adjustment a little off so it makes it look like the ship is going faster than it is."

"Oh yeah, I like that," Gerry said. "But ... wait just a tic. Yeah, now that I kick it around, I think the one-tenth light speed would be about right for how far they still need to go and make it here when they say. My gosh! They said they had already slowed way down, and they're still going thirty thousand kilometers per second. What if this is real?"

"Too wild mate," Simon said. "We'll just need to hang tight until the radar guys decide to kick it again."

The news of this anomaly soon filtered through international networks and reached an operations officer in the Situation Room of the White House at 3:42 a.m.. Since the message described it as probably a false contact, the operations officer decided not to make the wakeup call to the president.

By this time, the Copernicus Space Telescope had picked up a definite bright spot at the coordinates. This telescope, like its predecessor the Webb, was designed to work into the near infrared range, but it also had a substantially larger mirror and a greater capability in the visible wavelengths than the Webb. Whatever object caused the image, it shown brightly in the visible wavelengths. Everyone at NASA Mission Control knew there should be nothing like that in that area. But they couldn't think of any possible equipment problem that would cause that image. They hadn't heard of the radar echo yet and had no idea how far this object might be. Barely moments after their initial puzzlement, they heard that the Extremely Large Telescope—ELT and the Very Large Telescope—VLT of the European Southern Observatory, both in Chile, had also picked up the image. In South Africa, it was dawn, and the Southern African Large Telescope would need to wait through another day before it could join the hunt. But with two ground-based confirmations, no one doubted the reflection was real.

As speculation about the first radar echo circulated widely, scientists at the Southern Hemisphere Planetary Radar System sent a second pulse and planned to send another once every hour. After the long wait for the second echo to return, they began receiving a regular echo each hour still coming in at the same 91.9 degree angle from the planetary plane. All showed that whatever it might be, it was traveling toward Earth at one-tenth of the speed of light. No one now thought that it was an error, but no one could understand how their instruments could be picking up a space vessel at that distance. Others also thought of the transponder idea but had dismissed it since the echoes seemed genuine. By the time fourteen hours had passed since receiving the first echo, the object had traveled toward Earth to a distance 3.1 billion kilometers away—still farther away than to the orbit of Uranus. It was still traveling at one-tenth light speed and at the same 91.9 degree angle.

Washington, DC, The Situation Room, Saturday, 13 April, 5:00 p.m.

President Kaitland called an emergency meeting for a select number of experts from the EDS Oversight Group at 1700 Saturday. Oscar Wainwright, the Secretary of Defense, and several from the Joint Chiefs of Staff were also in attendance. The meeting was held in the White House Situation Room because of its advanced and secure communication equipment. The president wanted to receive new reports as quickly as possible. She had been studying all of the incoming reports. With her physics background, she had no difficulty understanding the technical issues that troubled the observers at the various facilities. The last thing she wanted was another mystery. But she was troubled. Almost everyone believed by then that Earth's radar was, indeed, tracking the alien vessel—what else could it be? Radar data showed that the object had a spherical shape and did not seem to be rotating. It should not be possible to detect such detail for something as small as a spaceship. The optical observatories too obtained images of a blurred but bright disc—again not the expectation for a space vessel at that distance.

She called the meeting to order and asked for ideas. She wanted to hear their ideas before sharing her own thoughts. Within a half hour, she'd heard the complete range of concerns—all ones she had already thought of. Everyone seemed to be on the same page with her. Everyone understood what the data were indicating, but no one could believe the alien vessel was so large that it would appear as it did on the images.

One hour into the meeting, another set of reports arrived. The staff officer handed the report summaries to President Kaitland. She read them in silence and frowned as her face went pale. Then she sighed and said only, "We need to discuss this latest information."

She handed them to the Commander of the Space Force on her right indicating that he should read them aloud.

As the Space Force Commander read the summary statements, faces around the table registered shock. General Beckworth's face blanched. Oscar Wainwright heard the commander read the estimated diameter of the spherical object plus or minus an uncertainty range and was shocked to hear that the number was scaled not in meters but in kilometers! He thought it must be a mistake, but no one else raised it. He was in his late sixties, the oldest person in the room. Even though the US had at last switched to the metric system in 2045, he still thought in the old way much of the time. He could calculate a rough conversion to miles in his head, but he wanted to confirm it. He picked up his phone, punched the kilometer figure into the calculator, and looked at the answer now in miles. He stared silently at the screen for a moment—object diameter = 612 ± 31 miles—*Sweet Jesus*!

Chapter 16

ANTICIPATION

Canberra, Sunday, 14 April, 9:00 a.m.

About one hour after that moment in Washington, Gerry was back at the Tidbinbilla complex. He and Simon were studying the latest reports they had snagged off their many com links. When Gerry saw the estimated diameter of the sphere, he felt a physical reaction as if something inside had snapped apart. He looked at Simon incredulously. Gerry said, "How can this be?"

"I know," Simon said. "It's comparable to some of the larger moons of Saturn—it falls into the size range of a dwarf planet. This is crazy. But how could all of our instruments be wrong? They've been checked and rechecked again and again."

Gerry shuffled through the reports with no real hope or plan. He said, "I can't think what to do except wait for a few more reports to come in and see if there's any change. It's still heading toward us at a tenth of light speed and will have to slow down soon."

"It had better be the alien vessel," Simon replied. "If a mystery dwarf planet that size came out of nowhere and impacted us at that speed, Earth's mantle would be vaporized—or worse if you can imagine that."

"Yeah, it has to be the aliens, and like you said, it better be. We'll be getting new reports soon," Gerry said. "It's still a long way out—still out a lot farther than Saturn's orbit. Maybe we'll get something new to bite into. When people hear about this, some will panic." Gerry said goodbye and told Simon he'd be back in the evening.

Simon had been spending all of his time at the complex for weeks. He was something of an introvert but was glad to hear that Gerry would be back. "Do you suppose you could bring a large pizza when you come back?" Simon asked.

"No doubt about it, my friend," Gerry said as he went out the door. He headed home to meet Sandra who was finishing an early shift at the hospital. She had caught a rough one dealing with accidents coming out of the late Saturday night and early Sunday morning hours—people drinking and driving. Gerry and Sandra had a nice brunch and early afternoon together before she fell asleep.

At half past seven that evening, Gerry walked up to the door of the control room where Simon had set up his temporary abode. Laden with ten family-size pizza boxes plus wearing a large backpack stuffed with assorted snacks and drinks, he struggled to open the door. He thought it would be nice to share the feast with Simon's colleagues too. Everyone there knew Gerry well, and he was always welcome—particularly that evening with his banquet. They settled into the treats while talking over the latest reports. At 8:00 p.m. a report arrived that showed the object to be only a little farther away than Saturn's orbit. The optical images were clearer, and there was greater certainty about the sphere's diameter. Its angle and speed remained the same.

Gerry said. "The size of this vessel is staggering. The UN invitation instructed the aliens to assume an orbit one hundred thousand kilometers from Earth's surface. That sphere had better be mostly hollow or it could have a strong enough gravity of its own to mess up things like Earth's geostationary and geosynchronous satellites. I wonder if that's been bothering anyone else. I wonder if the experts,

ANTICIPATION

wherever these decisions are made, are thinking maybe they should tell the aliens to stay farther out."

Gerry and Simon continued musing about such things until eleven that night when fatigue called a timeout. Gerry left for home, and Simon headed for his cot still happy with his memory of the pizza.

Washington, DC, Sunday, 14 April, 9:00 a.m.

About that moment on the other side of the world, President Kaitland called another meeting in the Situation Room with the same group as the evening before. Reports received overnight indicated that the object had reached a point as close as Saturn's orbit at about 3:00 a.m. Washington time—that time back corrected to account for the 1.4 hours it took for the radar echo to reach Earth. If the vessel maintained its present speed, it was expected to reach a point as close as Jupiter's orbit at about 9:30 a.m. Washington time. A radar echo confirming that location should arrive at Earth 44 minutes later at 10:14 a.m. President Kaitland expected there was enough to talk about that the group would still be meeting when they received that confirmation. At the time the meeting started, the most recent reports showed a much clearer image of a featureless disc and a tighter estimate of its size. It was now reported as 986 ± 9 kilometers in diameter. It's speed and angle of approach remained the same.

President Kaitland sensed the tension in the room. Not a single face looked anything but sober and troubled. She went first to the newest addition to the group, Mikalos Vanecek, a Lieutenant Colonel in the Space Force who also happened to have a PhD in astrophysics with a specialty in planetary systems and with a great deal of research experience.

The president said, "Colonel Vanecek, would you bring us up to date with what you gathered from the most recent reports from the observatories?"

"Thank you, Madam President," he replied. "First, the aliens left their solar orbit and traveled a substantial distance out of the planetary plane toward the Southern Celestial Hemisphere. Their mid-journey message came in at a 7.4 degree angle off the ecliptic which told us that—"

"Excuse me, Colonel, what do you mean by the ecliptic?" This question was from one of the group's few members who was not well-versed in astronomical terms.

"Oh, sorry," said Vanecek. "The ecliptic is only another name that can be used for the planetary plane. The name originated long ago when astronomers worked out that a solar eclipse is only possible when the Moon crosses that plane at the same time as it happens to be directly between the Earth and the Sun. I'll try to avoid specialized jargon.

"So, from the message's 7.4 degree angle off that plane at mid-course, we can calculate that the vessel had moved about one hundred AUs out from the plane."

"Excuse me, Colonel," said the same man innocent of astronomical terms. "How far is that?"

"Sorry again," Vanecek said. "An AU, or astronomical unit, is just the average distance of the Earth from the Sun—a handy measuring unit in the Solar System. One hundred AUs is a long way. It's three and a third times as far out as Neptune's orbit—the planet farthest from the Sun. And here I'm only using Neptune's orbit as a handy reference distance. Their ship is not close to any of the planets—it's that same great distance away but far south of the planetary plane."

Vanecek continued, "So being way out there, we presume they paralleled the planetary plane inward toward the Sun for the greatest part of their journey. We think they did that in order to go around the Kuiper Belt and get into cleaner space that has much less Solar System rubble. They could travel faster there. The next part is a bit puzzling. The most recent message came in at an angle of 91.9 degrees from the ecliptic." Vanecek sent an illustration from his phone onto one of the room's display screens. It showed a sketch of the planetary plane and a colored line showing the ship's estimated path. "The best we can figure is that they went past Earth inward while still a hundred AUs out from the ecliptic to a point abreast of the Sun and then turned ninety degrees aiming directly toward the Sun for a time.

"This is conjecture, but it appears that after a period of travel toward the Sun, they began angling back toward Earth at that 91.9 degree angle from the plane. The only reason we can guess why they did that was to obtain an overall view of the Solar System from that viewpoint out of the planetary plane and opposite the center of the Solar System. Maybe it was a special survey of some kind. In any case, they are now approaching Earth on a straight course from the southern polar sky.

"Second, the size of the object is as surprising as it is disturbing. We now have a more dependable estimate of its diameter, and it's certain it's in the size range of a dwarf planet. I'm concerned about its mass. What will happen when the vessel assumes its assigned orbit at 100,000 kilometers from the Earth's surface? Will the vessel's own gravity interfere with our geostationary satellites posted at about 35,800 kilometers above the surface? We already know that the Moon's gravity does have a slight effect on those satellites. The Moon is much bigger than the aliens' ship, but it's also about four times as far away as the alien vessel would be.

"At this stage I can only make a rough guess of the ship's mass based on its size and an assumption that it is substantially hollow. I referenced Earth's largest supertankers as one guide to the mass of structural metal used for volume enclosed and adjusted for an advantage of the spherical shape of the alien ship. I used the density of titanium as another guide assuming that titanium might comprise a major portion of its structure. We know nothing of its internal arrangement or what it might be carrying. In guessing what fraction of its volume is hollow and what fraction is structure and equipment, I chose a wide range that I expect should give an upper limit to its mass. I arrived at a range of about 1.1 to 8.9×10^{17} metric tons."

Surprised murmurs rippled around the table.

He went on. "Using that mass range, I calculated the vessel's gravitational force on the geostationary satellites when in closest alignment and compared it to the force of the Moon's gravity when it's in alignment the same way. The vessel's gravitational force on the satellites would be about 4 percent as strong as the Moon's force at the lower end of the vessel's mass range and about 33 percent for the top estimated vessel's mass. In my opinion it is more likely to be near the lower figure.

"It's good that the UN specified an equatorial orbit for the vessel. That eliminates transverse forces on the geostationary satellites which also are in equatorial orbits. But the added force from the aliens' ship might still have a small effect on east-west drift of their positions. As I said, the effect of the Moon's gravity on the satellites is slight and the ship's effect would be even less for the mass range I have estimated. Each of the satellites is equipped with 'station-keeping' thrusters with propellant used to correct each satellite's position from time to time. The station-keeping thrusters will be able to compensate, but they will consume propellant in doing so.

"When the aliens are close enough, we can ask them the precise mass of their vessel, and they might tell us. Alternatively, after they enter Earth orbit, we will be able to judge its effect. I don't expect a serious problem unless the vessel's mass is much greater than my upper estimate. Considering all those points, I would like to throw the question out to the group as to whether we should ask the aliens to take up a more distant orbit from Earth as a precaution." He fell silent.

"Thank you, Colonel Vanecek, for that comprehensive summary," President Kaitland said. "You've cleared up a number of uncertainties that were in our minds. We have other important things to discuss here this morning, but why don't we begin with your question. Would anyone like to jump in?"

General Beckworth was the first to gain her attention. "Thank you, Madam President. I'm grateful to Colonel Vanecek for this information about the vessel's mass. I was worried about that too but didn't work out an estimate like he has. I'm glad to hear its effect should be less than the Moon's. We can double-check with the UN's International Telecommunication Union that manages the satellites' positioning. I'm guessing that the vessel's effect will be acceptably small.

"But even if we knew there would be a larger problem, I believe we must accept the aliens into their assigned orbit. The entire logic for the orbital positions of the EDS missiles depends on the alien vessel being in that assigned orbit or very near it. If we persuaded the UN to have the aliens park far enough away to eliminate gravity problems from the ship, then all of our effort to create the EDS system would be in vain. Yes, we could shuffle all of our missiles to new orbits farther out, but that would need to be done in full view of the aliens—surely a provocative act. But more important, the one hundred thousand kilometer orbit is already a stretch for us—it's near the limit of where we can have any confidence that our defense strategy could work. If the aliens are significantly farther out, our EDS is just window dressing to make the public feel better."

"Thank you, General Beckworth," the president said. "That makes it very clear—it's a critical point. Is there anyone here who has a counterargument to General Beckworth's assessment?" A number said they strongly agreed with him while the rest nodded assent.

"Good," President Kaitland said. "I believe I've heard enough on this question. I agree with General Beckworth too. Thank you again, Colonel Vanecek. It was an excellent update for us and an important point to bring up."

She continued. "Let's hear the latest update on the status of the deployed defense missiles. Colonel Garcia, would you help us with that?"

"Yes, Madam President," he said. "Four days ago there was concern whether we would make it, but two days ago the Russians completed deployment of the nine missiles they had promised. And yesterday, the Chinese finished deploying their promised group of five. That now gives us fourteen from the US and Europe and fourteen from the Russians and the Chinese. We would like to have more, but twenty-eight missiles spread according to the logic of our strategy experts give us a chance of success if we need to use them. The Russian missiles are armed with larger warheads, a worry in the past but now a benefit. The personnel at missile command centers for all participating nations have been trained in the agreed rules of engagement. We would like to have had extra time for more intensive training, but it is ongoing and will be a continuous exercise even after the aliens arrive. All missiles are armed, fueled, and prepared for launch from their orbital stations on command."

"Thank you Colonel. It's a great relief that our Russian and Chinese partners came through as promised." The President began to fill the group in on news from the State Department concerning how Earth would receive and converse with these new ambassadors for whom there was no established protocol. The problem had become very complex with many nations jockeying for a position of prominence and little agreement being reached so far.

The communications officer interrupted the discussion with the latest observatory reports that just arrived. He handed the print to President Kaitland. She summarized it out loud to the group before passing it around. She said, "The latest radar probes show that the vessel is now as close to us as Jupiter's orbit, and the angle from the plane has remained the same. Its speed toward Earth is still constant at thirty thousand kilometers per second. The radar data and the latest optical images yield a tighter estimate of its diameter now judged to be 987 ± 5 kilometers. If the vessel maintains its current speed, it will reach as close to us as Mars' orbit by midafternoon." Then she added, "From memory, Mars' orbit is only about four-and-a-half light-minutes away from us." Then after a pause and with a touch of black humor, "I do hope they know how to slow down."

The meeting ended, and there was silence as the report was handed around. If a mindreader had been present, each mind in the room would have presented troubled thoughts—anxiety in one of its many disguises for such, strong disciplined people. General Beckworth remembered that the aliens' earlier position beyond the Kuiper Belt was 1,540 times farther out from the sun than Earth. The aliens took only a few weeks to cover that distance traveling at nowhere near their top cruise speed and even including a detour way out south of the ecliptic plane. He also remembered how difficult it was for Earth even to traverse the Solar System like with the New Horizons Pluto probe that set a speed record taking eight years and eight months to reach Neptune's orbit from Earth. The aliens at their much reduced "suburb speed," had gone that distance in less than forty-two hours and in a spaceship the size of a dwarf planet no less. General Beckworth was not alone in feeling what inhabitants of an ancient walled city under siege might have felt huddled anxiously while listening to the hoof beats of history outside their city gate.

Although some national governments preferred that the regular observatory reports from the Southern Hemisphere should be kept secret and classified, that preference soon proved impossible to achieve. Each report summary flashed around Earth's connected scientific communication networks and then leaked everywhere including onto broadcast news media. When the size of the alien vessel was revealed, panic struck numerous people who had already been afraid of the aliens. But it didn't cause broad movements of crowds into the streets. No one knew what to do other than cower indoors in dread.

Tucson, Sunday, 14 April, 2:00 p.m.

In Tucson that afternoon, Jim and Ellen heard the report confirming that the alien vessel was then as close to Earth as the orbit of Mars. About an hour later, they heard their phone sound an incoming call. Gerry and Sandra, both just out of bed and with cups of coffee already in hand, were as eager to talk about the report as Jim and Ellen were.

Gerry said, "Well, I heard that at the latest distance, they were still traveling at one tenth of light speed. They're only a half AU away now. Here in the south we might be able to see them with a pair of binoculars if we look in the right place."

"I'm really excited to see them with my own eyes," Sandra said. "But at the same time, I ... well ... I don't really know how to describe my worry ... I guess I should just own up to it as fear."

"I know," Ellen said. "It's the same with me. There's just nothing in my experience to equip me for this."

"There's nothing to compare with this—nothing this big," replied Sandra. "Nothing with such big consequences for good or bad. I'm trying hard to hang on to my optimism, but ..."

Jim said quietly, "You both just expressed exactly what I feel."

Serious for once, Gerry said, "Yeah, me too. I know I kid around a lot, but we all know the reality of what could be. We're just trying to stay in a positive frame of mind."

There was a pause all of them respected.

Then Sandra said, "So it's coming in from the direction of the South Pole, and the news has been saying things like it's as far away as Jupiter or as far away as Mars. But it's nowhere near the planets, so do you think they're saying that just as a guide because people have a rough idea of how far away the planets are?"

"Yes, I think that's exactly it," Gerry answered.

"Well anyway," Sandra went on, "they are coming close now. For the last few weeks, I've often thought about their arrival while I worked in the Emergency Room—I wondered—if things went terribly wrong, how would we take care of all the people who might be harmed by some kind of war? Maybe it won't matter. Maybe there's nothing we could do."

"I worry too," Gerry said. "We'll just stay together."

"Where you are it's already the fifteenth," Jim said. "It's the big day. Have you heard any other information about just when they will be arriving? I suppose you'll be able to see them coming as they get close."

"Definitely," Gerry said. "When they're close enough to spot, I know the precise place to look in the sky. Sandra and I will be out at Tidbinbilla later today with Simon. It's far enough from Canberra to have dark skies, and Simon has a nice telescope that we'll be using to watch that patch of the night.

"As to when they might be close enough to start settling into their orbit around Earth, I think what will happen next will be like those big transcontinental trains that go flat out for most of their trip, and then on the edge of their destination city, they slow way down and take forever to crawl through all the suburbs before pulling into the main central station. They are finally close enough that they will need to begin slowing down soon. My guess is that it will already be getting dark in Canberra by the time they arrive. We'll head out to Tidbinbilla by midafternoon with plenty of food to share, and we plan to stay the night if necessary. They said they would arrive on the fifteenth, but maybe they meant the fifteenth in Europe or America, who knows. We'll see by tonight."

Jim said, "I wish Ellen and I could be there with you. It would be great if we could all be sitting around together watching this. I don't know when we'll first see them here, but since they're taking up an equatorial orbit, I'm sure we'll see them once they're settled into it. Maybe we'll need to wait for night here though; I don't know."

"Oh, I've done a few calculations. You two will see them whether it's night or day," Gerry said. "It will be big in our sky—even a little bigger than the Moon. It's hard to imagine how that will look with two moons in the sky. I remember way back wondering whether I would even be able to see the Visitors' ship with the naked eye except as a tiny spot of light. How things have changed.

"I've also worked out how long it will take them to make one swing around Earth in their orbit. They'll circle the Earth approximately every 95 hours, 54 minutes and 54.4 seconds," Gerry said proudly.

"Approximately?" Sandra chuckled and then muttered something about Mr. Spock.

Gerry laughed too and said, "Well OK—that's five minutes shy of four days, so hey, let's just call it four days. Everyone on Earth will get to see them regularly. There will be another moon in the sky that goes through all of its phases every four days instead of once a month."

The other three digested this for a moment picturing it in their minds. Then Sandra said, "OK, we need to start packing the food and comforts so we can head for Tidbinbilla later and maybe a night of sky watching. I'm glad I had off today and tonight!"

Gerry added, "We'll call you later today with anything new. Bye for now."

Washington, DC, Sunday, 14 April, 6:00 p.m.

President Kaitland was meeting with people from the State Department and those involved with the UN. All were only too aware that six hours later at midnight, the date would shift to the fifteenth. The latest radar reports confirmed that the vessel had slowed to a fraction of its former speed as it approached Earth. Estimates were that it would arrive in its assigned orbital position by the next morning Washington time, the hour and minute uncertain depending on the ship's speed in its final approach.

The radar and optical telescope data now confirmed the ship's diameter to be very close to 988.5 kilometers. It would be impossible to miss seeing it overhead. Colonel Vanecek had left a written report with the president that morning explaining that the alien ship would appear in the sky as 3 percent larger than a so-called "Super Moon"—the apparent size of the Moon when it is at its closest orbital approach to Earth.

For everyone at the meeting, this was uncharted territory. There was no single acknowledged representative to speak for Earth although the UN had been allowed that role when issuing the invitation. Now several nations did not want to concede that distinction to the UN and were attempting to push their way to prominence.

President Kaitland convened the meeting and said, "I know that we are all frustrated that this issue is not yet settled, and we all know it's not for lack of trying. But here we are still facing it. Secretary Varma, would you please bring us up to date on the latest negotiations?"

"Thank you, Madam President," the secretary of state responded. "It's been clear for a long while now that no nation with a shred of self-importance wants to stand by meekly and allow some other nation to speak for them to the aliens. China and India, with the size of their respective populations, each believes that it should be the first to greet the aliens and welcome them as honored visitors. The US, and by that I mean a consensus of the US Congress and a handful of recent national polls, claim the US should have the honor. They think the US is the most deserving because of its history of space exploration, its status as one of the world's biggest economies and because it's still a superpower. Russia is claiming a superpower and a spacefaring status too and demanding the same thing. Oh, and I should mention that China and India are adding the spacefaring angle to their claims as well. These are the main stubborn players. Britain and separately the European Union had pushed earlier to have the honor, but recently they have joined nearly all other nations in voting to hand the distinction to the United Nations. That is our present impasse. When the alien vessel arrives, numerous voices will reach out to it with greetings as if from the host. It might not be so bad as long as the words are well chosen, but if someone wants to act like a big shot and suggest they are more than they are, it could be embarrassing for Earth and confusing for the aliens."

"Thank you, Secretary Varma," the president said. "It appears the situation is no better than in our last report. I've been thinking about this. First, I'm confident that the aliens will not be confused. I believe they know all of us far too well to fall for some phony claim. And maybe I'm too much of an optimist, but I expect that some of the posturing you have described will fall away when the moment comes and some of these nations will be sensible. So my preference is that we join with the bulk of nations in a greeting from the UN. I believe in the long run that will be more beneficial. But, in addition, I suppose as a concession to Congress and the polls, it would be fine for us to send a short personal greeting as well. I suggest something like—'Greetings to you from the People of the United States of America—one of Earth's free and independent nations. We welcome you in friendship as our honored guests.' Maybe we can come up with smoother words, but please keep it simple and sensible."

The meeting went on for another hour while in Earth's southern sky far below Washington's southern horizon, the bright disc perceptibly grew in size.

Few Executive Branch officials and military personnel associated with the Earth Defense Shield slept in the hours leading up to midnight on April 14. Most expected to be awake all night and hoped they would still have clear heads the next day.

Chapter 17

ARRIVAL

Canberra, Monday, 15 April, 2:00 p.m.

As the seconds ticked past midnight in Washington DC foreshadowing April 15's Arrival, it was 2:00 p.m. on the fifteenth at the Tidbinbilla Deep Space Complex southwest of Canberra. Gerry and Sandra had just arrived there, earlier than first planned, to stay with their friend Simon the rest of the day and perhaps the night too. When they walked into the control room carrying bags of prepared food and various goodies, Simon looked up and said, "Terrific—you're here! Wait 'til you see what I have to show you."

They managed to get in their "hellos" and put down their bags before he led them outside to the east side of the building. He said, "It wasn't so good earlier, but now the sun is beginning to move to the west, and it's a little better."

They went around a corner and into a shady nook of the building that was open to the southeastern sky. There he had set up his own personal telescope. He said, "We can see it already without the scope, but it's much better with it. Later on tonight, the scope should let us see some details. Have a look."

The aliens were still coming in on a straight course, and the scope's clock tracking mechanism could keep their ship in the field of view reasonably well. Sandra and Gerry jumped at the chance to look. While they took turns, Simon told them that the latest radar data showed it had slowed a lot and was now about four times the distance of the Moon away. Radar was not showing much on its surface; only some small irregular features no one could interpret. The Copernicus Space Telescope had shown a faint hexagonal pattern over its surface suggesting a massive underlying support framework in the construction of its hull.

Back in the control room, they looked at radar images on the monitor. Simon also had downloaded the latest optical images from the Copernicus Space Telescope and the Southern Hemisphere optical observatories. He had plenty of information to share with them, and four hours went by without notice. When they returned to the telescope, each carried a chair to settle in for the duration of the approach. The Sun was settled into the western sky but still more than an hour from sunset. As soon as they rounded the building corner near the telescope, they saw the ship easily visible and shining brilliantly in the southeastern sky. They looked at each other silently knowing that each would always remember where they were and what they were doing at this moment in history.

Washington, DC, Monday, 15 April, 3:06 a.m.

President Kaitland and her select group of advisors had assembled in the Situation Room of the White House. Each had attempted a two-hour nap just after midnight with imperfect results. They were consuming coffee while they studied the latest reports. Under the tension of the last months, they had developed a strong sense of rapport, and each had confidence in the others' abilities. The latest reports showed the vessel at about twenty thousand kilometers farther out than the Moon's orbit. It was moving toward Earth slowly enough to reach its orbital position in two to three hours. The Moon had been full two days earlier and still looked nearly full shining brightly in Washington's night sky. But the vessel could not yet be seen in the North American sky since it was still below the southern horizon and would be for a while longer until it moved closer to its equatorial orbit. For now the president and her advisors were watching the vessel's image on monitors fed by NASA's partners in the Southern Hemisphere. The vessel's surface was still a mystery. Although it did have a few small surface irregularities, they had no discernable purpose. Otherwise it appeared smooth showing only a faint hexagonal pattern.

By 3:06 a.m. the Visitors' ship had approached another twenty thousand kilometers and was the same distance away as the Moon's orbit. That may have been the benchmark the aliens were observing, or there might have been another reason why at that moment the frequencies assigned to them sprang to life. Everyone in the Situation Room was startled by the sudden sound of the gentle alien voice.

"Dear People of Earth, we thank you for your kind invitation to visit you. We are now approaching your planet and aim to settle into our assigned orbit soon. We realize that the size of our vessel might alarm you. We assure you that it will cause no harm at the assigned distance. We have anticipated that your scientists would be concerned about our vessel's mass, so we will provide some details and explain why there is no need for worry. Our vessel's diameter is 988.74 kilometers, but its mass is only 1.00767×10^{17} metric tons. The gravitational effect of our mass on your high geostationary satellites will be very slight. However, since our vessel will pass around your planet about every four days, it could eventually cause a small amount of drift off ideal position for those satellites. We have the means to correct that drift by providing tiny, very precisely controlled impulses to the satellites. We will do that while we are here so that you will not need to expend valuable propellant for your station-keeping thrusters.

"We also apologize for the intrusive appearance of our vessel in your sky. We have the means to change the reflectivity of its outer surface and will now reduce its albedo to match that of your Moon. We will continue sending regular reports on our position as we approach. Once we are in our stable orbit, we look forward to beginning a happy visit and to having many pleasant conversations with you. Once again, thank you for your invitation."

And sure enough, as those in the Situation Room watched on the monitors, the extreme brilliance of the shining alien vessel dimmed until it looked like another moon in the sky but without the familiar "man in the moon" face.

There was silence in the Situation Room for a moment, and then Rachel Kaitland said, "It's difficult to know where to begin."

Tucson, Monday, 15 April, 1:40 a.m.

Jim and Ellen had managed a nap after Gerry at last quit talking at about 11:00 p.m. Tucson time. Now they were awake and on the line again to Tidbinbilla. They had listened to the Visitors' broadcast and watched the vessel's reflection dim on their TV screen. The great majority of humans on the planet were also awake and watching with whatever contrivance they could reach. Human emotions around the world ranged from extreme to extreme. Some people were on their knees praying fervently, others were trembling on the edge of panic and still others were overwhelmed with awe. The four friends also experienced flashes of all those emotions playing back and forth in their minds. For those people who were most fearful, the human emotion of hope still persisted if feebly. Even a condemned prisoner standing before a firing squad still hopes for a last-second reprieve.

At Tidbinbilla, Gerry, Sandra, and Simon watched the glowing disc slowly grow in size and shift from its southeasterly position toward the north to assume its equatorial orbit. Its image in the telescope was stunning now, but its surface detail was as mysterious as ever. Gerry was no fool, and he well understood the breadth of possibilities that might soon eventuate. But that sober part of his mind yielded to the curious child part that was wild with excitement. On the video, Jim and Ellen easily related to that child part of him.

Gerry said, "These guys are just too much! Not only were they considerate enough to dim the reflection of their ship—and who knows how they did that—but they also took care of our worries about the mass and gravity of their vessel. And my gosh, why does it have to be so big? How could they need a ship that big for an exploration vessel? Sure, their trip is lasting an incredible length of time, but still it seems crazy-huge unless they're all giants. I didn't get that from their description of themselves."

"I know! I've been thinking the same way," Jim said. "The size of their ship—it's stupefying! And another thing—it's so immense, isn't there some danger it could collapse into itself under its own gravity? After all, it can't be solid. The mass they gave us for it is so enormous that I can't help but wonder."

"Hey Jim, that's a good question," Gerry said. "It hit me too when thinking about its scale. I was wondering how it would have to be built to handle the force of its own gravity and its acceleration when it moves. I was able to dig out some facts I needed from the computer here and have made a start."

"Great! Have you worked anything out yet?" Jim asked.

"Well," Gerry said, "first I worked out that its surface gravity would only be about three one-thousandths of Earth's surface gravity. That's pretty low and confirms most of it is hollow. But it's so huge; it would need gigantic bracing beams and all kinds of internal structure to hold its shape against acceleration. And no ordinary beams could provide rigidity over the immense spans of length needed. I'm guessing it's done with extremely clever bracing and maybe some kind of cellular structure. I imagine they used low-density, high-strength materials like titanium, graphene, and high-pressure foamed metals. Also I'm assuming there's an internal pressurized atmosphere that would help stabilize it like an inflated ball. But still there must be the risk of immense resonance waves in a structure that big."

He paused ... Then, "You know, an image just came to me—an imaginary encounter of early Pacific Island colonizers crossing the Pacific on their log and grass-rope-tied rafts meeting one of our biggest modern supertankers in midocean. They would look up at it in awe and wonder how it could ever be possible to build such a vast ship. I guess all I can say is the Visitors must have solved all those problems somehow. Maybe they even have some kind of force field that's related to their ship's mysterious drive system—who knows? Anyway it's a real thing, and it's right here before our eyes."

Then Gerry launched into a long description of how he worked out an estimate for how much of the vessel's mass would be structure and how much would be "stuff" that it carried like engines, aliens, and their suitcases. He proudly announced that the vessel would be able to carry a cargo of about half its total mass, namely 50,000 trillion metric tons.

Jim made a mental note to be careful about what he asked in the future.

And by then, Ellen and Sandra were having a power nap.

Washington DC, Monday, 15 April, 5 a.m.

In the Situation Room, the president and her group followed details of the approach through their many communication links. The Visitors had been broadcasting regular reports of their position and velocity, and Earth's radar systems provided simultaneous confirmation. There had been little discussion or comment by the group while they watched this momentous drama unfold. Not a single person in the room was unaware of just how historic these moments were in the entire human pageant. All were lost in thought about what might happen next.

President Kaitland had an abundance of worries such as what might go wrong with their defenses including the reactions of the Russian and Chinese partners. She also wondered when would be the right moment to send the welcome message from the US. There had been no prearranged precise moment that the UN would send its greeting, but she thought the US should send its message right after the UN's. *And how many other nations would send individual greetings?* She hoped all would choose appropriate and dignified language that would not cause embarrassment to Earth.

General Beckworth's mind was also troubled. His intellect informed him that everything so far about the aliens had been a hallmark of courtesy and consideration. All of their communications had signaled benign friendliness and even helpfulness. But his entire career as a defense strategist tuned his mind to prod his emotions with every barb of deceit and betrayal of trust that he had experienced or read about. And now the aliens' colossal vessel was arriving perhaps with weapons so advanced that humans couldn't understand them and maybe with an army large enough to conquer the planet. Everything about the aliens suggested stupendous power compared to what Earth could muster. How could they reach down more than sixty thousand kilometers and nudge the geostationary satellites? Would the Earth Defense Shield prove only to be a foolish and feeble waste? As he watched the screens, he wondered how he could know if the aliens were about to begin a surprise attack. After all, surprise is the operative word! He was strong and disciplined and forced his mind to stay focused. That worked in the main, but sometimes, like a fatigued muscle, it could falter.

By 6:00 a.m. Washington time the immense alien ship had approached its equatorial position closely enough to become visible to Northern Hemisphere viewers in the Pacific. It slowly entered its orbital position to revolve around Earth in the same direction as the Moon— west to east. And in what seemed a curiously consistent affection for the International Date Line, the great ship settled into its assigned orbit directly above it one minute before midnight there Universal Coordinated Time. The Visitors said they would arrive on April 15th, and, indeed, they did. They always adhered to Universal Coordinated Time (UTC). By mid-April, cities like London and Washington had gone onto daylight savings time and cities like Auckland and Canberra had gone off it. The Visitors' official arrival time was recorded for Auckland at 11:59 p.m.—for Canberra at 9:59 p.m.—for London at 12:59 p.m.—for Washington, DC at 7:59 a.m.—and for Honolulu at 1:59 a.m.—all on the fifteenth.

Tucson, Monday, 15 April, 5 a.m.

An hour before that moment, Ellen and Jim said goodbye to Sandra and Gerry to go outside to watch. Dawn was not far away, and the Moon, just two days past being full, shone brightly in the western sky. They were confident that people in Hawaii could see the ship and those on California's west coast could too, but they weren't sure about their location farther east and behind the mountains west of Tucson. It was an indescribable thrill for them after they stepped into their backyard and saw the great shining sphere edge up just over the western mountain ridge as it nestled into its final stable orbit. It definitely appeared larger than the Moon and, like the Moon, was almost fully illuminated. Seeing two great moons in the sky for the first time filled Jim and Ellen with a kind of primal awe that they could not have articulated. Understanding what they were seeing only added to the import of the moment.

But it was a brief view—their ride on Earth's turning surface was already carrying them away from this newest moon sending it back down below the western mountain ridge to be followed soon by Earth's old moon. Brief though it was, the first view of a twice moonlit night was still a once-in-a-lifetime experience they would never forget.

Jim thought way back to New Year's Eve when they had watched the celebration on Sydney Harbour. It now felt so long ago, and despite all the many troubles of the world back then, that time seemed to him naive and innocent compared to now. They stood there a long while with their arms around each other watching the dawn light strengthen and remembering the twin moons that heralded a new stage in the human story.

Chapter 18

NEW MOON

Within moments of the Visitors' ship being settled into its stable orbit and a brief announcement from them to that effect, everyone on Earth had the same idea. Messages of greeting poured out from every country. It had already been arranged that no matter where the Visitors settled into orbit, line-of-sight transmitters would be available to send collected messages from all sides of the planet. The UN message was a dignified and welcoming greeting. The US message President Kaitland had suggested, now polished and refined, was also sent. Russia, China, the European powers—all did the same. Later, when the messages were published in the media, President Kaitland was relieved to find that none should have caused Earth embarrassment. Humanity had lifted itself up to do justice to this extraordinary event.

But in the hour following the intense excitement of The Arrival, a feeling of anticlimax began to settle in. People wondered, "Now what?"

The Visitors had broadcast a brief but warmly worded thank you message to all of Earth for its welcoming greetings but then went quiet. Only twenty minutes of silence were needed for some people to begin worrying that something frightening might be about to happen.

High government officials wondered what they should do. Who will talk to the aliens first? Will the UN need to be the principal voice of Earth or should each government try to establish their own diplomatic relationship? Why were the aliens staying silent? Were they too busy closing down their ship's drive systems after the long journey? Didn't they have someone in charge of communications who could handle this? For all the knowledge that the Visitors had about Earth's people, they still did not fully appreciate just how excitable and fearful human beings could be. And for humans, they still needed to learn that beings having such a long life span as the aliens are more patient.

After a few hours, fears and confusion were calmed when the Visitors' frequencies again came to life. One almost had to wonder whether the Visitors were showing off because now they were broadcasting their message not just in the eleven languages of the past months but in eighteen of the Earth's most commonly spoken languages. And as usual, all were perfectly written on the screen and beautifully pronounced by the gentle voice.

"Dear People of Earth, we thank you again for your invitation to come here and for your warm and gracious words of welcome. We anticipate that you will have many new questions to ask us, and we are happy that now we will be able to talk with you much more easily. Our ship is now only one-third of a light-second away; how delightful that will be for conversation. We understand that there might be uncertainty about how to talk with us. For the present, if you continue to send us questions over the usual frequencies, we will receive them. We will prepare a more responsive communication system over the next several days and will notify you when it's operational. We also understand that many of your national governments wish to have direct communication channels with us. We will be able to develop such a system soon, but we must emphasize that we do not favor one government over another, one nation over another or one person over another. Please understand that, from our point of view, you are one people.

"During our journey here, we gathered a great deal of current information about significant bodies orbiting your star out through the radius of the main Kuiper Belt. We merged this new information with a similar survey we completed when we first arrived in your system. From that merged data, we have now been able to prepare a catalog of all potentially hazardous asteroid-like bodies and comet nuclei that might develop a collision course with Earth. We have also included optimal ways for you to protect yourselves for each situation using your current technologies. But this catalog will not be fully reliable for more than about 150 of your years because the random odd shapes and varying compositions of potentially interacting bodies in the outer Solar System make it impossible to calculate precise collision rebound vectors. It will still be necessary for you to be watchful. Nevertheless, we hope you will find this information helpful.

"We will soon resume the regular information broadcasts that we had been sending before our journey here. In addition, we are adding an extra broadcast channel that will send a continuous view of Earth from our orbital position along with added data communication links for current information relating to weather, ocean conditions, and land features. We hope you will enjoy the view.

"We will sign off while we prepare our new communication systems over the next few days. Goodbye for now."

The voice went silent and in a few minutes, the promised informational broadcast began. Earth's planetary scientists were overjoyed with the new data on possible future collision objects. The view of Earth from the ship's position was spectacular in ultrahigh-definition, and as scientists learned later, the broadcast signal also contained image data in additional wavelength providing even more valuable information.

Earth's people were calmed and reassured by the message. In the hours following it and with the resumption of regular broadcasts, people began to relax and believe that the Visitors' arrival had brought no new threat, at least not for the time being. They began to think of new questions to ask, and before long relay transmitter stations were receiving thousands of questions to send to the ship above.

During those first hours for those able to see it, the immense vessel created a new and stunning spectacle. It moved in the same direction as Earth's surface, but because Earth spun around four times for each revolution of the ship, the vessel's apparent movement in the sky from a view on Earth's surface was westward like the Sun and Moon. In the first hours while Earth turned under it, it passed over Australia and eastern Asia. Through that first night there, it provided a twin-moon view to all who looked up. At times later when it was overhead during daylight hours, the Sun could not wash its splendor from the sky even as its changing phase diminished it to a half or to a crescent moon. Earth's people struggled to picture its position and motion in their minds. Toy manufacturers and computer coders soon set about creating models to demonstrate the dynamics of the relative motions of the Earth, Moon, and the Visitors' ship. Business was brisk.

Over the course of the next four days, the Visitors' vessel made one complete revolution around Earth giving everyone a chance to see it. It was now universally dubbed in the media as "the Ship." Seeing it as a real object in the here and now had a strong effect on many people and prompted them to think about new things in new ways. The internet was jammed with searches related to astronomy, physics, alien life forms, and the interface between religion and science. Governments were relieved that the public was remaining orderly and still attending their jobs and obeying laws. During that first four-day revolution, the Visitors remained quiet with only a few brief updates relating to new contact frequencies for particular kinds of communication. It was a time of questioning and reflection for most of Earth's people.

Many wondered when the Visitors would be ready to begin having conversations with Earth and, indeed, what they meant by those words. Were they still collecting and sorting through more new questions? No one had a sensible idea of how conversations would be carried out and with whom. No one had an idea of how all those millions of earlier questions could be addressed in any meaningful way.

Tucson, 15-19 April

During those four days after The Arrival, Ellen and Jim had a chance to recover from a missed night of sleep. They kept up a regular video link to Australia talking about their own questions and feelings.

Gerry's enthusiasm was again running at the top of the scale while Sandra tried to keep him down to a reasonable level of excitement during video calls. On one of the calls he said, "You know, Jim, I've been thinking again about the Ship being so big. You're right—it's freaking crazy big! I've been making some calculations."

"Hey, good Gerry!" Sandra interjected. "This has been driving me crazy too."

"Yeah, this is nice stuff," he said looking pleased at Sandra. "So I've been thinking—how would they have structured the inside of the Ship? First off, they're bound to use different measurement units than we do, and I'd guess they use a different numbering system too—one with a different base than our base-ten system. So everything I've played with here will be wrong in detail but might be on the right track in concept. I'm guessing they put the drive engines, nuclear reactors, and extra nuclear fuel at the core of the sphere. That way any waste heat from them would radiate outward toward the outer shell and help warm the interior while they travel through deep interstellar space where it's near absolute zero. So with that kind of stuff in a massive core structure, I imagine that immense struts would radiate outward from the core in all directions to support the outer shell.

"It's near the outer shell where I'm guessing their living and working quarters are. I have no idea how much space they might use that way so I just took a shot to see how much space could be available. I imagined a series of spherical decks going inward from the outer protective hull. I figured the outer hull shell must be pretty strong. I guessed it at four kilometers thick and made out of an interlocking network of struts with an inner and outer metal shell. Inside that strut network would be room for pipe runs, wiring, and hangars for things like excursion vehicles, antennae, and similar things needed near the surface. So then down at the inner shell of the hull, you're already four kilometers into the sphere.

"Next I dropped down two full kilometers to the first deck. Hey, it's a big hotel, and I like the idea of a full two kilometers of 'sky space' over each deck. I imagined each deck's layer to be two hundred meters thick for structural strength. It would be filled with strong structural girders between an upper and lower shell but still mainly hollow to accommodate plumbing, wiring, transport tubes, and all that stuff. So that means that the open space above a deck uses up two kilometers of the sphere's radius and the deck itself uses another 0.2 kilometers so that's a total of 2.2 kilometers for each deck layer. Now imagine a total of thirty decks like that repeating inward and using sixty-six kilometers of the Ship's outer radius. It's like one of those Russian dolls—thirty spherical decks each a little smaller and nestled inside the next larger one. So standing on any one of those decks, you could look up to see two full kilometers of open space reaching up to the bottom of the next deck above. The only things you might see to the sides would be one of the massive main struts radiating from the core and some occasional lesser struts to support the layers."

Sandra said, "Gerry, this is awfully elaborate. How can you know any of this?"

"Honey," Gerry said, "I'm just having some fun with it. I just want to get an idea of how much usable space might exist in a ship that size. Just hang with me another minute. So where was I—oh yes, what proportion of the total sphere's volume would those thirty decks occupy? Well it turns out to be only a little more than one-third of the Ship's total volume! Man, that leaves a huge amount of space inside for all that core machinery stuff and lord knows what else they might have."

"Whew," Ellen laughed. "This is quite a shaggy dog story, Gerry."

"I know, sorry," he said. "But I think you will like where I'm going."

Jim said, "Yeah, keep going Gerry. I'm really into this—I love it!"

"OK," Gerry said. "So, the next question is how much area are we talking about for those decks? Each of them has a generous amount of sky space overhead so you wouldn't feel claustrophobic, but just how big are they? This called for spreadsheet time to work out the surface area of each. First off, how big do you think the area of the outer surface of the Ship is?

"I have a feeling you will tell us," Sandra chuckled.

"Hmpf!!" he smiled. "Drumroll, please—the area of the outer surface alone is greater than the area of Alaska, Texas, California, and Minnesota combined."

"What? Get outta here!" Jim exclaimed.

"No, I'm serious," Gerry said. "Grab your calculator and have a go." Then he said, "And how big is the first deck inward? Naturally it's smaller than the outer surface. But still it's as big as Alaska, Texas, California, and Georgia all added together. Heck, even deck number thirty, the smallest innermost one, is still bigger than Alaska and California combined. I added up all thirty decks for fun, and it's crazy. Their total combined surface area is greater than all of Asia and Africa added together—the Earth's two largest continents. Man, their area is ten times larger than the whole of Australia!"

"You have got to be kidding us!" Ellen said.

"No!" Gerry said. "This is all for real. And I can do one even better. This is the *pièce de resistance*! What if I had decided to give them only one full kilometer of sky space above each deck? That still allows enough sky space to be way more than the tallest skyscrapers on Earth. If I had done it that way and doubled the number of decks in that outer third, the total area for those sixty decks would add up to more than all of the land area of the entire Earth including Antarctica—it would exceed Earth's total land area by as much as the whole area of Australia!"

"I'm speechless!" said Jim.

"Me too!" said Ellen.

Sandra jumped in with, "This reminds me of how the internal surface area of our two lungs is as big as half a singles tennis court."

"Right, honey—spot on! That's good," Gerry said. "This is all good fun just to appreciate how big it really is. But it still doesn't tell us why it needs to be so enormous. I still can't figure that out."

"Me either," said Jim. "And you know, there's another thing bothering me. While you were describing these decks, I was imagining them standing on each deck and imagining scenes like we might see here on Earth. But then I realized they would have no significant gravity unless they have invented artificial gravity which seems quite a stretch. And I remember from our radar data that the ship was not spinning while it approached us so unless they have separate spinning sections inside, they're not simulating gravity that way. That method would be awkward anyway for a spherical structure."

"Yeah, Jim," Gerry said. "That's a great question to bring up. The gravity from the ship's mass itself would be puny like I mentioned the other day. I've been wondering about that too. I remember they said that they liked the freedom of living in space. They also said they had worked out how to protect their bodies from degrading in the absence of gravity. I wonder if part of that freedom they love is freedom from the confining effects of gravity. Maybe they have become accustomed to floating everywhere. We're so accustomed to thinking of up and down and always being pulled down toward the Earth that it's ingrained in us. When I was working out the arrangement of those decks, I was thinking in familiar scenes like you mentioned. But on the ship, it might be nothing like that. I can't guess how they might have everything laid out if there's no up or down. Man, how I would love to have a guided tour!"

"If you figure out how to get one, be sure to ask for tickets for us too," Jim said.

Then Gerry said, "I'll bet you thought I was done."

Sandra groaned.

"There's more," he said, "but I'll make this one quick. First off let's go back to the two kilometer deck spacing with only thirty decks. Remember that every bit of this is indoor space even if it is bigger than Earth's two largest continents. I wondered what I could do with that. So I thought, let's give every person on Earth a big mansion on a very generous lot—say a five-hundred-square-meter house on a half-hectare estate. So one of these goes to every single living human being on Earth, newborns to retirees—let's just say ten billion. Of course I've included plenty of space for roads, parks, city centers, and natural landscapes. I'll spare you all the numbers. But after the calculator finished, the thirty decks would accommodate every bit of those ten billion estates with parks, roads, etc. and with several billion square kilometers to spare." Gerry couldn't stop himself once he got going. "What I'm saying is that a lot of aliens could be living in that ship. Some people might find that a little frightening."

"I might be one of those people," Sandra said quietly.

On another one of those chat sessions, Gerry started talking about eclipses. "You all realize" he said, "that we'll have more solar eclipses now."

There was silence—no general chorus of—sure, we know that.

"Yeah, OK then," Gerry said pressing on. "The Ship crosses the ecliptic plane every two days because its orbit is above Earth's equator and because of Earth's tilted spin axis. When it happens to cross the plane at a time that it is directly between the Earth and Sun, a solar eclipse will occur somewhere on Earth. And it will be a good one too since the Ship is a little bigger in the sky than the Moon and will completely cover the Sun. But it won't last as long as one with the Moon since the Ship moves around the Earth in four days instead of a month. I haven't heard that anyone has calculated a schedule yet for when they'll occur, but I'll bet someone is working on it."

The other three loved to tease Gerry, but they did enjoy these jolly excursions from him. Sometimes they even had the chance to talk about other things, but in those first days, Gerry's excitement mostly got the better of him.

Washington DC, Friday, 19 April

The bewildering size of the vessel came up in the first of President Kaitland's morning meetings with her group after the Arrival. Reasonable explanations for its size eluded all leaving only frightening speculations to fill the void. Colonel Vanecek, the planetary science expert, was another person who liked to play thought experiments and had worked out an interpretation of the Ship's possible internal structure that helped him visualize what it might be able to contain. When he explained his rough estimates to the rest of the group, they found even more reasons to worry. And that led to a discussion of whether the Earth Defense Shield would be a credible weapon against such a colossus. One person suggested that it only takes a tiny pinpoint to burst a balloon, but that analogy did not gain support. Nevertheless, all agreed that surely a direct nuclear burst on the hull's surface would penetrate through causing great internal damage. Several direct hits might damage it enough to cause the aliens to retreat or possibly end hostilities. So far nothing had happened in the remotest possible way that suggested the Visitors would be anything but peaceful and friendly. But the extravagance of their apparent power kept these analysts on edge.

During the next few days while the Ship completed its first revolution around the Earth, the Visitors continued their informational broadcasts with a few brief added announcements of new contact codes that the UN and national governments could use to communicate with them in a more direct way. President Kaitland found this a welcome addition, but all of her officials were still on tenterhooks. All governments involved with the Earth Defense Shield were continuously checking its status to be sure it remained operational. Ongoing training of the rules of engagement continued with all command personnel. Despite the tension, President Kaitland felt relieved that nothing threatening had happened in the first days after the Arrival. And around the world, a majority of people were still enchanted and awestruck by the spectacle in the sky.

In Washington on Friday morning, the Visitors' ship was two hours into its second revolution around the Earth. President Kaitland had just finished her regular morning meeting with her select group and had returned to her private study off the Oval Office to have a scheduled quiet hour to review two recent reports. It was rare during the day for her to have an opportunity for quiet concentration, and she was looking forward to it. She asked her personal secretary, Eleanor Atwood, to hold all phone calls unless it was an emergency. She had just settled into one of the reports and was making good headway when her mobile phone on the desk sounded. In the first second she felt annoyance but then alarm—*it must be an emergency*. She picked it up and answered expecting to hear Eleanor's voice even though Eleanor would normally have called through the main desk phone.

But she did not hear Eleanor's voice.

The voice said, "Hello, President Kaitland."

It was the gentle alien voice she had grown so accustomed to during the last three-and-a-half months.

Chapter 19

GETTING ACQUAINTED

Washington, DC, Friday, 19 April, 10:08 a.m.

For a few seconds Rachel Kaitland was stunned—disoriented. But she regained her senses enough to respond with. "Hello. Who is this? Did my secretary connect you through?"

"No," the voice answered. "I called you directly taking a chance that you might have a few minutes to talk with me. My name is Michael. I am one of the Visitors, and I am calling from the large ship that arrived at Earth four days ago. Have I called at an inconvenient time?"

No adequate words exist in any language to describe how unexpected these few sentences from this voice were.

She stammered, "Nu ... nu ... no ... I ... I do have some time available just now. But ... how ... how were you able to call this phone directly? All incoming calls must first pass through the building's main switch board and then through my secretary's switch board before being passed to me." As she said these words, she was beginning to wonder whether it was some kind of outrageous prank. How could it be real?

"We have a way of making direct contact," the alien voice said. "It is important that I talk with you. I apologize if I startled you. I hope I have not offended you."

213

"No, no," she said still almost stammering. "You haven't offended me. I was only surprised. Are you the same individual who has broadcast messages to us in the past? Your voice sounds the same."

"No, I am not the same person, and I must apologize for that confusion," Michael said. "We have used the same voice synthesizer profile for all of our messages after determining that it would be a pleasant-sounding voice to most Earthlings. I see that I made a mistake to use it just now. I will look into introducing a unique sonic element to each of our voices for me and other of my people who communicate directly with Earthlings from now on."

Rachel Kaitland's mind struggled to think rationally and pull away from the surreal feeling of this conversation. It was disorienting to be referred to as an "Earthling." She realized this might be the most important phone call of her life, and yet it had the casual relaxed quality and tenor of a new neighbor calling to begin an acquaintance. She knew she needed to ask the right questions, and they were not coming quickly to mind as usual. At last she thought to say, "Please understand that I want to ask you many questions. Please understand that we have many concerns that need to be discussed. Will I have an opportunity to speak with you again?"

"Oh yes; don't worry at all," replied Michael. "I will give you my direct contact code, and I will welcome your call any time of day or night. I don't need to sleep. I have been selected to be your full-time contact. You will have exclusive use of my time and should never hesitate to call me if you have any kind of question."

The intensity of Rachel's emotional response to Michael's reassurance surprised her. A powerful mixture of surprise, relief, gratitude, and anxiety washed over her consciousness in waves. She had no time to think ahead to future concerns over possible deception or misinformation from this new source. For now, she was equally astonished to be contacted this way and relieved to be given the means to communicate directly with an alien representative. She gathered herself and said, "Thank you. That is a great relief to me ... you mentioned a code. How will I use that?"

"It is easy to use," Michael answered. "I will give it to you now if you would like to record it. It is a series of four symbols and twelve numbers. To reach me, you need only to enter the sequence into your phone as you would do for any other phone number, and I will answer."

President Kaitland still felt as if she were teetering on the edge of a dream. But she listened and wrote down the code and repeated it back to Michael.

She asked, "How shall I address you? You gave me your name as Michael. Here that is a masculine name. There is so much yet that we need to learn about you. Is it appropriate that I think of you as masculine?"

Michael laughed gently.

She was startled and definitely pleased. It was her first inkling that the aliens might be capable of laughter.

Michael said, "Please don't worry that you might offend me. Your language does not have an appropriate pronoun for beings like me who are both masculine and feminine. It is fine for you to think of me as masculine and to use masculine pronouns when referring to me. And my name Michael is all that you ever need to address me."

"Thank you for explaining," Rachel said. "It will make our conversations easier for me. May I ask how you chose the name Michael? I assume that it is not a common name where you come from."

Michael again laughed gently. "Yes, you are right about that. I hope you will not feel that I have invaded your privacy by reading a great deal about you. I have read all of your published papers and all biographical information about you that is available. From that I learned that beginning in your youth, you have had a high regard for Michael Faraday, the renowned nineteenth-century scientist. I humbly admit to indulging in a small amount of manipulation in assuming his name."

This time it was Rachel who laughed. "No, I don't mind in the least. In fact, you could not have made a better choice." She found herself relaxing without caution. She felt like she did when she was much younger and just beginning to become acquainted with a fascinating new friend. Michael's voice and manner were so engaging that she almost forgot who she was and who she was talking with. But her sense of responsibility pulled her back to her job. The obvious question occupying everyone for days came into her mind, and she said, "I am very pleased to meet you Michael and glad that I will be able to talk with you regularly. I wonder if I might ask you a question right now, one that has puzzled me."

"Yes certainly," Michael answered. "Please do."

Rachel said, "You must realize that the physical scale of your ship is far far greater than anything we on Earth imagine as needed for a vessel of exploration. Would you explain why it needs to be so large? Does the drive engine require such a great size? Or are many of you needed on board to tend it? I don't know what information you are free to share with me, but I would appreciate whatever you can give me. All of us here are very curious." She did not voice General Beckworth's concern that it looked big enough to house several invading armies.

Michael did not hesitate. He said, "Oh yes, we did realize that you would be curious about that and perhaps frightened too; we expected to be asked. There are some questions I will not be able to answer for you, but this is one that I can. The ship is so large because our journeys are long, and there is much about our home region that is difficult for us to leave behind. Long ago our ships were much more modest, and our journeys were shorter. But over time as our technology advanced and our resource base expanded into nearby uninhabited star systems, energy and materials became so abundant, that we developed ships that provided approximations of 'the comforts of home' so to speak. We are also social beings who take great pleasure in each other's company. There were many who wanted to join this expedition, and with our adventures so far, they are very glad they did. As big as this ship is, and it is one of our larger vessels, it is still a pale miniature of the many space habitats in our home systems. But it is comfortable. I would never complain." He added a small chuckle to that last remark.

Rachel, finding this somewhat hard to swallow, pressed on. "You said there were many who wanted to join the expedition. Is there then a large group of you on board?"

"Yes, it's a sizable group but not so many as to crowd the activity and entertainment decks of the ship. Most of us on board are scientists of different specialties but all with an interest in exploration. We also have historians, artists, poets, entertainers, and others. All together we form a vibrant and diverse community and keep ourselves busy with our main interests and with enjoying our friends.

"Oh yes, you were interested in the total number. Oh my, I can't recall the precise number in your numbering system. Please give me a moment … OK, yes here it is—all together we number 19 billion, 822 million, 472 thousand, and 324 individuals."

There was a sustained silence. If President Kaitland had been on a video screen with Michael, he would have seen her mouth fall open and her face turn pale.

After President Kaitland said goodbye to Michael, she had less than a half hour of her scheduled quiet time remaining. She immediately called Eleanor, and asked her if she had patched any calls through to her private study. Eleanor said no—none of the calls that came in were identified as emergencies so she had held them. The president spent the rest of her quiet time recovering from, thinking about, and writing notes of the conversation. She found herself dwelling on a few specific things. One was Michael's friendly and open nature; he did not seem to hide information. She also remembered that he needed to check on the exact number of the ship's population. It was good to see that he had at least a small limitation. But the main thing that kept recurring in front of her mind's eye was that number—the number she herself would remember—19,822,472,324. On board the Ship, there were roughly two aliens for every single living human being on Earth.

President Kaitland's reminder alarm let her know that her quiet time was finished, and within a few seconds, the main phone on her study desk sounded.

It was Eleanor. She said, "Madam President, I knew you would want to know right away. Since you checked with me a half hour ago, you have received five calls from members of your Select Oversight Group plus a question from Albert Thompson who came into my office to see if you were free. All of them said it was not an emergency, but all of them did want to talk with you quite urgently."

"Thank you for letting me know, Eleanor. Who was the first of those calls?"

"The first was Dr. Jonathan Schneider," Eleanor answered.

President Kaitland had given Jonathan Schneider a break from the operational meetings of the last several weeks but had asked him to rejoin the group when the Ship arrived. "I'll walk to Albert's office now to talk with him," Rachel said. "Would you please call Dr. Schneider and put the call through to me there?"

Albert's door was open; she entered and asked her chief of staff, "Albert, you wanted to talk with me?"

"Yes!" he said with excitement in his voice. "I'm glad you're here. You just won't believe this! I had a call a little while ago from someone claiming to be one of the aliens—she claimed she was calling from the Ship!"

"Join the club, Albert," the president said. Then the phone sounded. "It's probably the call I asked Eleanor to patch through. Put it on speaker please."

Jonathan Schneider sounded much more excited than usual. He said, "Madam President, I have news that you need to know about right away. I've been contacted by the alien ship."

President Kaitland had Eleanor contact the other four in quick succession and heard the same report from each. She asked Eleanor to cancel a less urgent meeting coming up soon and to arrange an emergency meeting with her Oversight Group. She only gave them

GETTING ACQUAINTED

enough time to drop what they were doing and make their way to the West Wing. Everyone arrived on time, and all were animated.

She convened the meeting and began, "I know that a number of you have important news, and I do as well. I will begin by saying that this morning I was contacted by phone by a voice sounding identical to the voice of the alien messages. The voice said it was one of the aliens on the Ship and was calling me directly from it. Now, has anyone else here had a similar experience?"

Six people raised their hands, and the others stared with astonished expressions on their faces.

She said, "Would each of you who were contacted first give us a short description of what happened and what was discussed. Then we will delve into issues that seem most urgent to examine."

The short summaries were very similar to President Kaitland's experience in the first part of each call but varied in the later part as each person regained their composure and thought of particular questions to ask. It happened that the last person she came to was the only one who had the presence of mind to record the conversation—the commendably capable Colonel Vanecek. Everyone in the room was eager to hear the recording. President Kaitland asked Colonel Vanecek to link his phone to the room's sound system. Colonel Vanecek did not think to start the recording until twenty-three seconds into the conversation when the caller gave his name. The recording began with the same gentle voice of the alien.

"... yes, I do have a name—it is Anton. As you would guess, I have chosen an Earth-style name to use here. My name in my native language would be unpronounceable for you as it is partly composed of acoustic compression waves like you use among yourselves, but also mixed with the addition of electromagnetic field modulations that we can both generate and sense."

This small bit of casually dropped new information produced an audible murmur of surprise from the listeners. The voice of the speaker was unmistakably the same as the gentle voice of the messages.

Colonel Vanecek said, "Thank you for using a name I can pronounce. That is very helpful. May I address you as Anton? Please feel free to call me by my first name, Mikalos."

"Thank you, Mikalos. Yes, Anton is fine for me too," said the voice.

The conversation went on for several minutes with the exchange of introductory pleasantries until Colonel Vanecek took the opportunity to gain some key answers. He explained that he had been very puzzled by the immense scale of their ship and wondered why it needed to be that large. He asked Anton whether many of his people were needed to operate such a large vessel. Anton answered that the ship mainly ran itself with what Earthlings would call an integrated artificial intelligence system. Colonel Vanecek asked again why the ship was so large.

Anton answered, "The planned journey was long. We are beings who greatly value the benefits of companionship and community. We did not always have the ability to build on this large scale, but for more than thirty of your millennia we have been able to do that. Now before beginning a long journey, we recruit people who are interested in exploring with us, and we typically wind up with a large enough group for us to feel at home with our expedition community. It's very much more pleasant for us that way."

Colonel Vanecek said, "That's very interesting, Anton. What a delightful way to go on an expedition if you have the means to take along that many. May I ask how many of your people are with you on your ship?"

"Certainly," Anton replied. "It's nothing like the communities we have at home, but it's still a good group. Including myself, the total number of us on board is 19 billion, 822 million, 472 thousand, and 324 enthusiastic travelers."

On the recording there was a loud thump-like sound as of something dropping on a floor—presumably on Colonel Vanecek's side of the conversation. There were also distinct gasps from the listeners around the table.

After a pause that included a few more fumbling sounds on the recording, Colonel Vanecek responded a little shakily, "That ... that is ... that is, indeed, a very good group. I am guessing that you know enough about Earthlings to know that we would find that to be a quite large number to travel on an exploration vessel. I'm surprised that even your vast ship is big enough for so many. I wonder—is it possible that you are smaller than humans? Would it be impolite to ask you how large you are? You know, we are fascinated by you and very curious to learn more about you."

Anton laughed good naturedly. Those around the table who had not received calls were surprised to hear alien laughter for the first time.

"No, I don't mind at all. I'm happy to explain," he said. "Yes, we now are smaller than humans. In our earlier naturally evolved state long ago, we were larger than you, approximately twice your average mass. You might recall from one of our earlier answers that we came from a planet with a greater surface gravity than Earth's. Although one might think that would lead to a smaller, lighter form, it encouraged a stout body structure to support our mass and to better compete with other vigorous life forms. But much later after we learned how our genetic code directed the embryonic development of our bodies, we were able to choose and modify how we wanted to be.

"It was a gradual process of many small steps with each significant change being carefully considered and agreed upon. Now we are much smaller. We have retained our six legs at the base of our bodies and our six arms at the top, and we can still move around easily in moderate gravity and manipulate smaller items with our arms and fingers. Our brain is still located in the trunk of our body between our arms and legs, but everything is smaller. We were even able to reduce the mass of our brain slightly by increasing its neuronal packing density somewhat like your birds have evolved on Earth.

"Now the average mass of our biological body is about twenty-seven kilograms. In addition, each of us always wears a close-fitting auxiliary machine pod that fits around our central trunk and has a mass of nearly ten kilograms. We control everything about our pods with the electromagnetic field modulations we generate with our thoughts, and the pod itself feels like part of our own body. The pod has a number of important auxiliary senses and manipulative devices that easily make up for our reduced body size. And very important, the pod structure fits precisely into our larger machines as a control module. When we 'wear' one of the larger machines—for example, like a machine to fly over the surface of a planet, or one to travel from one planet to another one nearby, or say, a large construction machine for assembling space habitats or vessels like our ship—the larger machine also feels to us like part of our body. Was that a helpful explanation, Mikalos?"

All the listeners around the table sat wide-eyed and staggered at this wealth of fascinating and gratuitous detail.

Colonel Vanecek answered on the recording, "Oh yes, Anton! This is wonderful! Thank you very much. This is very enjoyable for me, and I'm looking forward to more conversations with you. You mentioned that your home communities are even larger. May I also ask how many of your people are living in your home region?"

"Yes, of course," Anton said. "We even surprise ourselves at how large our community has grown in the past since the Great Understanding. The number of our people now exceeds ... let me think ... oh ... I'd say a bit more than two-hundred-and-thirty trillion."

Very audible gasps now came from all listeners around the table save Colonel Vanecek.

President Kaitland asked Colonel Vanecek to pause his recording. She said, "Colonel Vanecek, thank you so very much for thinking to record this conversation. I'm almost speechless over how important all of this new information is. And thank you for pressing on with these particular questions that give us such valuable insight into what we are dealing with. I have a question for you. Did Anton give you a special code so that you would be able to call him again for further questions and conversations?"

"He did, Madam President," Vanecek replied. "He was extremely friendly and helpful and invited me to call him any time I liked day or night. He said he didn't need to sleep."

"Thank you, Colonel" she said. "That is the same as my experience this morning with my call. And, indeed, I was quoted exactly the same number of aliens on board their ship. Unless I am mistaken, we will have the opportunity to gain much more information this way in coming days. And I expect that those of you who have not yet received a call, will receive one soon.

"Instead of our going back around to examine the calls further as I first suggested, would each of you who have had a call please settle down somewhere quiet and record every scrap of information you can remember before it fades or you are distracted from hearing other accounts. I will do the same. Once finished, we will have the notes compiled and printed for all to read. I'll have Eleanor arrange to have a stenographer available for each of you right away. Please do not delay. Every remembered point might be important. For those who may yet receive a call, please please remember to record it. I will adjourn this meeting for now. Please return here at 1700 for more discussion."

Chapter 20

NEW NEIGHBORS

Earth, 19 April to 18 July

If President Kaitland had hoped to keep the events of the morning quiet at least until her group had time to discuss their implications, that hope was quickly shattered. By early afternoon, media news everywhere was full of reports from people who said they had received calls from aliens aboard the Ship. The Visitors were not limiting their contact to high government officials. People from every walk of life were reporting long conversations and new facts learned about the Visitors.

Until it was common knowledge, almost every Earthling who received one of the early calls wanted to know how many aliens were on board the Ship. Soon everyone knew the number and the number of Visitors that lived back in their home region. Earth's people reacted to those numbers with indignant disbelief or awe and apprehension.

Astronomers who were contacted managed to extract more details related to the size of the home population claimed by the Visitors. They wanted to know whether the Visitors' home star would show signs of a 'Dyson Swarm'—the existence of so many orbiting habitats and starlight collectors that Earth telescopes would see evidence of their star's light being partly blocked by all the objects circling around it. The Visitors answered that they were spread out so evenly and in so many different orbital planes around a number of nearby star systems that even their trillions were not yet enough to cause an observable effect from that distance.

Others like Vanecek asked what the aliens looked like and how large they were. People were fascinated by the shape of their bodies. One intriguing new fact surfacing was that the Visitors had three fingers at the end of each of their six arms. And the predictable fact also came out that the Visitors used a base-eighteen numbering system instead of humans' base-ten system.

Conversations with biologists and medical doctors revealed more about the Visitors endoskeletons. People had difficulty visualizing how long bones like those in human arms and legs could work efficiently with six arms and six legs arranged around a circle. The media reported that afternoon that the Visitors' bodies were based on six lobes arranged radially around the long axis of the body's trunk. Each lobe extended upward out of the trunk to become an arm and downward from the trunk to become a leg. Each of the lobes contained a central bony but flexible spinelike structure that ran from the end of the arm above, through the trunk and to the end of the leg below. The skeletal spinelike structure was surrounded by masses of muscle—the aliens' version of muscle—and was more robust in the leg and the trunk region but finer and more flexible in the arms. At the end of each arm, the flexible bony core divided into three yet finer and more flexible spinelike cores for the three fingers. As a result, the Visitors' arms and legs were extremely flexible, but each could have the strength of a python.

More anatomical answers described a ring-like bony structure analogous to the human pelvis that connected the six lobes at the top of the trunk, and another stronger one at the bottom of the trunk above the legs. The Visitors also explained that their brains each had six lobes and all were contained together in their version of a bony skull just under the bone connecting-ring of the upper trunk. The Visitors even revealed that the six lobes of their brain were connected in a way similar to how the two lobes of the human brain are connected with the corpus callosum.

Biologists welcomed these fascinating details, but the average person wanted photos. So far the Visitors had not supplied images of themselves. Once the phone calls began, impertinent humans wasted no time in asking for pictures. When asked, however, the Visitors deftly managed to deflect the question with distractions and new interesting subjects. A consensus grew on Earth that the Visitors feared humans would find their appearance to be frightening or even repulsive and deliberately avoided sending any photos. Artists' interpretations based on how the Visitors described themselves continued to most resemble something like a sea anemone. Few people were satisfied with that.

By the time President Kaitland's second meeting convened at 1700, four others of the group had received calls from their now personal alien contacts. All four remembered to record their calls. The pattern was the same. Each Visitor caller had been extremely courteous and friendly, each had chosen the name of someone that the recipient respected or loved, and each had given the recipient a unique private code to call back at any time. The president asked that the audio files of the calls be copied to the White House system and asked Eleanor to arrange a transcript of each one. Then they examined the body of new information that had accumulated.

President Kaitland began, "If what we have been told is true, we are facing a situation more formidable than we would ever have imagined. What we see with our own eyes above us and what we have seen that they can do in traveling here supports their credibility. And at the same time, some of what we have heard strains our credibility. Does anyone else here think the idea of almost twenty billion aliens on board the Ship seems absurd? Could they all possibly fit even if each of them is only half the size of a human? Would their combined weight alone be too much? I haven't had time yet to try working through any calculations. Has anyone else?"

The admirably reliable Colonel Vanecek answered first. "Yes, Madam President. After I outlined my first ideas the other day about the Ship's structure, I did more detailed calculations because I had the same question in mind as you. I found that an arrangement of spacious internal decks, each with a half-kilometer of space between them, quickly added up to an amazing total area. The calculations showed that less than a fourth of the total volume of the Ship was enough to contain a combined deck area greater than all of the land area on the entire surface of the Earth even including Antarctica."

There was stunned silence and open-mouthed stares from all around the table. Then there were murmurs and one louder protest of frank disbelief. "Surely that can't be right!"

"Yes, it's incredible I know," said Colonel Vanecek. "And this morning after I heard how many aliens were on board and learned the mass of each plus the mass of their auxiliary pods, I added all of them together and compared their combined mass to the total mass of the Ship. It turns out that all of them together including their pods comprise less than eight one-billionths of the mass of the Ship—actually it's 7.278×10^{-9}. It surprised me too. The stupendous scale of the Ship truly is difficult to comprehend. They all can fit easily within it and have plenty of spare room to get lost in too."

After a short pause while everyone pictured those numbers in their minds, President Kaitland said, "Thank you, Colonel Vanecek, yet again for having thought it through so well. That certainly makes it clear. Well then, has anyone else wondered about the credibility of what we are hearing? Yes, General Beckworth?"

"Thank you, Madam President," General Beckworth said. "I must say I'm feeling neglected because I haven't yet had a call. I hope I will soon as I have many questions. But for now, I will say that the aliens have surprised me. I did not expect them to reach out to individuals the way they are doing, and it occurs to me that it is a brilliant strategy for reassuring a nervous populace. I too wondered about the claimed number of individuals on board. Almost twenty billion seemed absurd to me. And then another thought occurred to me. Their technology seems so far ahead of ours. I remember that the caller to Colonel Vanecek mentioned a highly integrated artificial intelligence system that runs the Ship. We can assume that AI system is far beyond anything we have seen. Now I wonder whether all the calls that are being received—and I might add, received by means we haven't worked out yet—I'm wondering whether these calls all originate not from individual living aliens but from a massive artificial intelligence system. And after hearing Colonel Vanecek—my God! What else might the other three-quarters of the Ship contain? What we thought we were inviting here was not this. What other surprises are there? Whether we are facing nearly twenty billion aliens or a massive artificial intelligence system with the power to simulate that many at once, this has become the most dire of challenges."

"I couldn't agree more, General Beckworth," the president said. "Yes, Dr. Schneider, please go ahead."

"Thank you, Madam President. I'm as astonished as we all are by these new revelations. And I agree with much of what I've heard this afternoon except that we may not be remembering well enough that the Visitors are, indeed, aliens—some things they do may seem absurd to us but be done for perfectly good reasons of their own. They speak our languages perfectly using idiomatic expressions and laughing at appropriate times. But is that behavior an accurate facsimile of how they really are—of how they would behave with each other—or is it only their polished efforts to act like us so we will feel comfortable with them? Or as General Beckworth raised, is it their AI system speaking and guided by all the information they have gathered about us in the last 136 years?

"They or their AI system have been able to simulate us in such a masterful way that it is easy for us to forget that they might not be like us in every way. We ought to remind ourselves that these beings are quite possibly far more intelligent than we are. We ought to remind ourselves that we are the primitives here. Their society may have achieved a level of harmony and affectionate civility that we have scarcely dreamed of. It's possible that they all exist together in a warmth and communion that we are barely able to achieve at times within the closest of intimate family relationships. So while it sounds absurd to us to bring along twice Earth's population on a trip of such magnitude in order to have enough companionship, well maybe they are just different than we are in that way. Maybe they are just being aliens from 240-plus light years away with an incredibly ancient and unique civilization behind them. Maybe that's enough of a reason for them to exhibit some behaviors we can't understand."

The discussions continued along these lines for some time with Schneider attempting to broaden the thinking of the others. Far to the west, the colossal Ship drifted serenely and majestically across the sky—a second moon with Earth's own ancient Moon, each shedding their glow on the quiet Earth below.

◈

The personal calls from the Ship began shortly after its first complete revolution around the Earth. The Visitors realized on the first day of their calls that they needed to make their voices individually recognizable. They were quick to add a unique sonic element to each that humans could recognize yet which retained the quality of the original gentle voice of the earliest messages. From then on, no one had difficulty recognizing their own special Visitor.

During the next weeks, several observations about the Visitors became clear. They actually seemed to enjoy talking with humans. They were unfailingly good humored, and they seemed to be endlessly patient with people's questions, even ones asking for photos.

If the number of Visitors on board the ship was to be believed, every living human could have his or her own personal alien ambassador. The Visitors continued to reach out to the people of Earth apparently aiming to achieve that goal. Anyone with access to a mobile phone, and that was the vast majority, was fair game. The Visitors soon expanded their language mastery to twenty-three of Earth's most commonly spoken tongues plus a few smaller groups of closely related languages. Inhabitants even of remote parts of the world were receiving friendly calls. Communication experts quickly guessed that the Visitors were bypassing Earth's interconnected switching networks and were broadcasting an appropriate signal directly to phones they wished to reach.

From signal tracing, it was obvious that they had deployed seven additional relay satellites around their orbital path adding one every forty-five degrees on their first revolution, a tricky feat in its own right. By Earth standards, it seemed an extravagant number of relay points for their elevated orbit, but the Visitors never seemed troubled by frugality and did everything to a high standard. They had apparently worked out all the details necessary for proper Doppler adjustments, timing corrections, and the functional signals needed to create their own relay network that would allow them to dip into Earth's huge population of mobile phones at will. Their only operational penalty was a modestly increased lag time when conversing with someone on the far side of the Earth from their ship.

When they were questioned about these relay satellites, they apologized but explained that they had been eager to begin conversations with Earth's people and had understood that their assigned orbit was theirs to use. It would seem that in their comprehensive study of all of Earth's literature, they had at some point come across the old maxim—*It's easier to obtain forgiveness than to obtain permission.*

Over the previous two decades, mobile phones had been adapted for routine interactions with satellite relay systems. Surface relay towers still provided the shortest lag times in populated areas, but otherwise thanks to low-orbit satellite relays, people could find good service anywhere, even in the middle of the ocean. Earth's satellite relays had receiving antennae adequately sensitive for the weak transmitters of ordinary pocket phones. But the Visitors' relay satellites were one hundred times farther from Earth's surface than Earth's own relay satellites. The inverse square law disadvantaged the Visitors greatly. However, as telescopes were trained on their relay satellites, it was soon clear that the Visitors had built what was needed—each relay satellite had a multicomponent antenna array connected by thin struts and cables and spread out over an area of more than fifty-four square kilometers. Experts could not but think that the Visitors were listening to more than mobile phones alone.

As the number of daily phone calls between Earth's people and the Visitors mounted into the billions, some of Earth's repressive governments felt unsettled. It was a totally unexpected situation that any ordinary citizen was able to communicate directly with the aliens. Even more confronting to these governments was that miscellaneous citizens had private contact privileges with the Visitors that were equal to the most highly ranked national ambassador or even to a head of state. Several such regimes found this practice to be an unacceptable liberty. But jamming attempts by those governments interfered with their own official use of their networks and was found to be ineffective in any case. The Visitors seemed to be able somehow to circumvent it. Confiscating phones was another solution, but enough phones escaped seizure that information from the Visitors leaked out to those repressed populations despite their governments' efforts.

Elsewhere around the world, after people heard from others that talking with a Visitor was like talking with an old friend, practically no one was so anxious that they hung up when hearing the Visitor's voice on their phone for the first time. Everyone who was contacted was given a personal code and invited to call back at any time. There was no shortage of people accepting that invitation. During those first weeks, there was a continuous frenzy of sensational media coverage of new information revealed daily by the Visitors. Only an extreme hermit could remain ignorant of what was happening.

The Visitors quickly showed themselves to be witty, funny, and extremely congenial conversationalists. Studying human cultures for the last 136 years enabled them to know humans very well. Perhaps because of their high levels of empathy and interest in humans, they were able to adapt to the wide variety of Earth's cultures and to individual personality differences. Their comprehensive knowledge of Earth's literature and history enabled them to converse on virtually any topic and even to entertain Earth people by recounting stories from Earth's own traditions.

Another surprise was how well the Visitors played with children. Many parents had not been able to resist bringing their children into a conversation with a Visitor or setting up whole family sessions with a Visitor. It was immediately obvious that the Visitors were delighted by children, perhaps by their unpredictability, spontaneity, and innocence of adult protocols. The Visitors were full of clever word games, knew every kid-joke ever created, and responded to children with great playfulness and endless patience.

For elderly people who had been living alone and experiencing a lonely existence, Visitor calls were literally a gift from above. They were able to talk with their Visitor contacts for hours at a time with no trace of resentment or impatience by the Visitor. The most trivial or repetitive stories and comments seemed to interest the Visitor contact greatly and were returned with encouraging responses. The Visitors even arranged a way to converse with people who had lost their hearing. They arranged a video feed to the phone screens with a human face speaking the words for lip-reading as well as scrolling text on the screen. There seemed to be no limit to their generosity and consideration.

It was not only fun and laughter. The Visitors were equally accomplished at giving serious and empathetic advice when the conversation called for it. People who had been troubled by depression, grief, trauma, or neuroses found the Visitors to be consoling and wise. The Visitors were soon recognized as excellent and supportive psychological counselors. Psychiatrists and psychologists were losing patients. Mental health professionals discovered that their own Visitor contacts possessed a profound wisdom about human motivations that astonished them. They began having professional discussions with their contacts and began incorporating the Visitors' psychological insights into their practices. A surprising number even chose to become de facto patients of their personal alien contacts.

Another unexpected effect was on dating behavior. People who were unusually shy or reticent with the opposite sex gave up for a time on their longing for a perfect mate in favor of conversation with their Visitor contacts. Others who typically had plenty of luck in love spent less time at that pursuit and instead favored conversations with their contacts. A bonus for both sexes was that all benefited from their Visitor counseling and felt much more assured about themselves and what they were really looking for in any future venture into romance. The phenomenon had become a favorite topic on media talk shows sometimes provoking hilarious humor and sometimes serious and thoughtful discussion.

The Visitors were so companionable that restaurants and entertainment venues suffered significant loss of business as people chose to stay at home and chat with their new alien friend. That is the term nearly all chose to use in referring to their contact. When talking with each other, humans referred to their contact as "my Visitor friend" and said things like, "My Visitor told me," or "What did your Visitor think about that?" As far as anyone could conclude from the Visitors' behavior, they were prepared to continue these conversations with Earth's people for as long as they were visiting Earth.

The first weeks of excitement and novelty passed quickly into two months of a relaxed level of comfort with the immense Ship overhead and ongoing conversations with the friendly Visitors. But some among Earth's people were not pleased with this turn of events. These included the indignant repressive governments and also several anti-Visitor groups who promoted an ideology they called "Humanity First." Members of these groups seemed to have developed a fulminating hatred of the Visitors born perhaps out of a feeling of human inferiority engendered by the Visitors' overwhelming presence. These anti-Visitor people persisted somehow in a kind of irrational and paradoxical denial of the Visitors' worth even in the face of irrefutable evidence. Others in the anti-Visitor camp simply regarded the Visitors as a threat to their privileged positions. The Visitors consistently and persuasively spoke of a society founded on empathy, fairness, nonviolence, and egalitarian harmony. Such concepts were anathema to some Earthlings, and anti-Visitor groups were beginning to organize and seek ways to counter the Visitors' growing influence.

They made first attempts with media campaigns claiming that none of the Visitor contacts were real individuals but were only synthetic fragments of the great artificial intelligence mind of the Ship—people were being duped and were only speaking to a computer. The anti-Visitor media campaign also claimed that the aliens were not interested in humans in a personal way but were cold-hearted scientists recording every conversation and playing word games with people in order to study human minds like a human scientist might study chimpanzees.

But these negative campaigns fell on deaf ears. People simply didn't care about these attacks on the Visitors' motives. They enjoyed their Visitor friends too much to be dislodged. And for a surprising number of people of faith, the attachment went even deeper. In their minds, the Visitors had obviously been sent by their deity as new teachers to help Earth turn away from its failings and become a better world.

◆

After so much effort preparing for disaster, the governments of Earth that created and maintained the Earth Defense Shield began to relax. They still maintained strict military discipline and a sense of vigilant readiness over the EDS system, but the tension and anxiety prevailing at the time of the Ship's arrival had given way to a steady state feeling of forces in balance. President Kaitland still held her meetings with the special advisory group but now only twice a week, barring emergencies. The meetings now had more the quality of routine status reports.

General Beckworth, always the most cautious and wary observer, received his first call from the Ship the day after President Kaitland's first call. His contact identified himself as Ian taking the name of Beckworth's favorite uncle who had been a strong role model and a dear and supportive companion in his early life. For the first two weeks Tom Beckworth grilled Ian, probing relentlessly for any inconsistency or hint of falseness. Ian remained unflappable and always responded without irritation, rancor, or any sign of having taken offense. General Beckworth took his job of protecting his country and his world very seriously, but he was a fair man. He could not find those fissures in the alien's answers that he had expected to uncover with sustained interrogation.

Before long he found himself looking forward to their talks for personal reasons. Living alone as he did, he had lengthy conversations with Ian nearly every day. Tom discovered that Ian was amazingly knowledgeable and insightful about Tom's favorite topics from Earth's science and history. After more than a month of these talks, their conversations began to drift into more personal aspects of Tom's life and eventually even into his memories of his wife and her death ten years earlier. Tom found it remarkable that he had come to trust Ian enough to venture into this still unsettled part of his mind.

After Tom's wife passed away from cancer following her long illness, his best friend Joe had been his strongest support and was always ready to talk or just listen. Other friends had been helpful too, but grief lingered taking up residence and refusing to go away. Friends offered comforting words—earnest and sincere and with the best of intentions. But always their words passed through his numb ears into the bleak landscape of his mind only to echo powerlessly off his memories of her.

Her doctors had not hidden what they foresaw, and for months before she died, they both knew she had very little time left. He struggled to find honest words of comfort and encouragement for her. For his sake she always put on a brave face, but that magnified his sorrow for her even more. Only one time did she admit that she hated to wake up each morning—wake to pass from a world of dreams into the real world of hopeless pain. They clung to each other that time finding a brief release in their embrace through the tremors of their sobs.

Despite her doctor's forewarning when the time came in the critical care ward, he was in no way prepared for that frozen instant of reality when, while he still whispered close into her ear how much he loved her, he heard the last rush of air escape her lips as her exhausted breast settled for the last time. And he was in no way prepared when his body fell upon her, and he wept piteously while the nurses there could not stop themselves weeping with him.

After that, other ordeals descended upon him. Barely restrained self-destructive impulses took over his life. In the past, he and Charlotte would sometimes sit together in the late evening in the dark and become absorbed in her favorite piece of music. She always wanted complete darkness to lose herself in the Vaughan Williams "Fantasia on a Theme by Thomas Tallis." After her death, he once tried to recapture that experience by listening to it alone, but it left him utterly shattered and aching in her absence. The nights were the worst when he lay awake and wrestled with the implacable searing knowledge that he would never again be able to speak with her and hear her answer. The finality of death took him by the throat and would not let go.

Through months later, a shocking and perverse part of his mind sometimes savaged him—mauling him with flashing images of her withered decaying face in her coffin or of her foot or hand decaying into its skeletal form. The long, tear-stained letter of goodbye he had placed in her hands just before the coffin was sealed appeared too and now only mocked him.

Tom surprised himself when he allowed these memories to surface in his talks with Ian. Somehow with Ian, in a way that Tom did not understand, trust was simply there. When he shared these memories, he did not really expect help from Ian. He would never have guessed that a being who had already lived more than fifty thousand years and who could expect another forty thousand years of healthful life would have anything helpful to say to him about human mortality. At an earlier stage in his life he might have laughingly compared it to getting marital counseling from a Catholic priest. But now he discovered that would be wrong. Wisdom can emerge in different ways from many directions.

Ian was limited to words through a telephone. His words began slowly, intermittently, gently slipping between Tom's memories like his wife slipping gracefully between the bedsheets beside him each night. They resurrected for him memories he had not yet rediscovered, some joyful, some jarring. They continued through session after session taking him each time to a peaceful mental meadow resting in sunshine. Ian's words were mesmerizing and at times morphed imperceptibly into a kind of poetry that he struggled to remember, but which always faded like a wonderful dream that on waking left him bathed in peace. Ian's words became for Tom a living, encompassing embrace that reached into his innermost core to heal. Tom understood then that the universe, always full of mystery, always ready with surprise, had worked some kind of miracle and had bestowed upon him a precious and true new friend.

Chapter 21

CHATS WITH SANDRA AND GRACE

The four friends also received their first calls and personal contact codes from the Ship. Sandra, Gerry, Ellen, and Jim continued their habit of regular discussions, but they now shared all the new things they were learning about the Visitors. Their contact Visitors had no reluctance in answering questions, even ones that Earth people would consider rather personal. The Visitors, in turn, did not hesitate to ask personal questions of the four friends. As with billions of other Earthlings, each came to trust their contact as an honest and well-meaning dear friend.

They also found that each Visitor contact had great expertise matched to each of the friends' professional fields. Ellen speculated that the Visitors were so advanced and, with their long lives, had accumulated so much knowledge that any of them would be an expert in any human specialty. She likened it to any competent human adult being able to help a beginning kindergartener with the first concepts of adding and subtracting. Jim thought that the Visitors might also self-select to match their own special interests to the interests of particular humans. But the friends learned quickly that their contacts would not hand over groundbreaking new discoveries. The Visitors provided helpful comments but insisted that Earthlings needed to make their own discoveries in order to grow into their new understanding authentically and to maintain their self-respect.

Sandra's contact chose the name Grace, the same name as Sandra's closest childhood friend. One day, Sandra asked Grace why the Visitors had embarked on a long process of deliberate genetic change. After raising the topic, Sandra said, "It must have felt quite risky to begin making significant changes to your body structure. There are so many interconnected and integrated systems in any complex organism that the chances of something unplanned happening are high."

Grace replied, "Yes, you are certainly right about that. But you would not be aware of how much experimentation and time we needed to go through before beginning to make these changes. Unlike natural selection, we did not simply try random changes and then let the failures die. We made progress slowly and carefully moving small steps at a time. If you think about the scientific progress your people have made in the past one hundred years, consider how much progress you might make in the next one or two thousand years. That future point for humans might be comparable to where we were when we began to make such significant changes in ourselves. Before that, we had already learned a great deal about the genetic control networks in embryonic development. We had already learned how to coach stem cells into growing a new healthy organ to be used for transplanting and how to grow a new eye in a damaged socket. We had learned how to stimulate the stump of a lost limb so that it would regenerate a new arm or a new leg. These were among the preliminary steps we mastered before moving on to larger goals.

"By the way, we have analyzed your genome carefully from all your studies and published sequences, and we see that you will be able to do exactly the same thing. It's only a matter of time until you work out the details."

"Oh my," Sandra replied, "thank you so much for that last comment! I have dreamed of these possibilities for years and am thrilled to hear that they could happen. And I see what you mean about patience in undertaking each step very carefully. I hope we will have the wisdom to manage that.

"Coming back to my first question, why did you decide to change your natural forms so dramatically?"

"Ah yes," Grace said. "That is a good question. As I think you have already heard, when we became sapient, our natural form was more than twice as massive as yours. Our form, like yours, had been refined over vast spans of time while running the gauntlet of natural selection. As I am sure you understand from your medical education, even such long refinement does not produce a perfect organism. The whole organism is just good enough to survive and reproduce effectively for the environment of its time. Natural selection leaves much unfinished business. Mutations are random, unplanned events. Nearly all are harmful, and 'Mother Nature', as you sometimes refer to the natural world, is ruthlessly uncompassionate and merciless. Those born hampered by harmful mutations die soon on their own or are taken by predators.

"Meanwhile, the survivors struggle on to play their part on Nature's stage with all of them unaware of the desperate insecurity of their lives. They may be fortunate enough to pass on their genes to offspring, but soon afterward they are taken away by an accident or a predator. They disappear and are replaced by others equally unaware and equally eager to pass on their genes.

"Well, we asked ourselves—could we improve on this process? We asked—is it hubris for us to take what we presently are, what has been so hard won over so much time of trial, suffering, and death, and imagine that we can improve upon it? In time with growing confidence in our knowledge, we decided that we could.

"The bodies we had then, in addition to various shortcomings, were much larger and more massive than needed for the kind of life we already were taking up. They were suitable for a primitive life on a planet with strong gravity and filled with dangerous predators. By then we were leaving our home planet and discovering that we liked the freedom of space. Except for our earliest ventures off planet, we soon built very livable, very comfortable space habitats. Certainly there were challenges. We needed to learn how to keep our health in low and zero gravity, and we needed to learn how to protect ourselves from the greater radiation present above our home world's atmosphere. We did learn those things surprisingly quickly and then went from strength to strength. This was when we began to seriously alter ourselves to a smaller body size and to develop machine interfaces for our technology. We call the people who began the process long ago the "Old Ones," and we honor them for their wisdom. We are descended from them but now so changed as to be a different species. But we are only different in form. We are their children."

Sandra asked, "Have you stopped with your self-transformation now or are you still continuing with a goal in mind?"

"We still are continuing the process of change," Grace replied, "but at a much slower pace now. A long-term goal we are considering is to eliminate our biological side entirely. Admittedly that is a very radical idea, and we have by no means settled on it yet. It won't happen unless we have developed our machine technology to such a degree that it is essentially like a living being but without the biological vulnerability we still possess. We would never go that far unless we were certain that our memories and cognitive abilities—our minds—could be transferred to a machine neural net with complete fidelity and that our minds would still authentically be our minds. We expect that this capability, if it is even possible, is still well in the future.

"But we have imagined how it might be if we reach that point. We have imagined how one person's mind and memories would be transferred to the new machine replacement and what would come next. We have imagined how that biological person would become like a parent to its machine replacement even though both would be exactly equivalent at that point. We have imagined how the two would become dearest friends with the machine person gradually becoming a caretaker for its biological parent until inevitably the biological parent finally dies. We imagine it would be much like the natural process of children taking care of their aging parents through to their inevitable end.

"But our greatest efforts at present are focused on extending the reliability and longevity of our biological bodies. Our life spans were much shorter in the time of the Old Ones, and only in the last half of our current era have we been making significant gains there. As you must have realized, our self-guided evolution could not have progressed as rapidly or at least in the same way if our life spans in the earlier stages of the process had been as long as we have achieved now. In the last sixty thousand of your years, our life spans have become much longer and our reproductive rate has become lower. For our people conceived in the middle or latter part of that period with the latest advances in their genomes, we don't know yet just how long their life spans might turn out to be."

Sandra was moved by Grace's deep feelings of respect and affection for the Old Ones and by her hard-edged description of nature itself. She felt much in common with Grace just then. She said, "I understand better now, thank you. The things you described to me—the facts about our biological vulnerability and nature's indifference to our fates—those are all things I have long thought about. Just now your words reminded me of a disturbing experience I once had many years ago. I was too young at the time for a visit to a private museum in the hospital where my mother was a doctor on staff. I was only sixteen and wandered into it while waiting for my mother to finish her work. It was a huge collection of babies born with terribly severe and fatal deformities and was there for medical students to study. Each baby was immersed in a large sealed glass jar of preservative liquid. As I wandered through the aisles of shelves, I was both fascinated and horrified. Eventually I fled weeping and was haunted for months by my memories of what I had seen. Since then I have seen even worse, and sometimes I am angry at the universe for its cruel indifference. I know that sounds ridiculous—I know it's completely irrational, but the feelings still come to me every time I see suffering."

Grace replied, "Oh my dear, dear friend, I hear tears in your voice even now, and I feel them too. You are not alone. And you are not being irrational. Yours is the perfectly rational feeling of a conscious living being looking into the stark reality of death and as far as we know, of nonexistence. Yours is the perfectly rational response to the unfathomable depths of impotent outrage that prevails throughout all the worlds of living things—outrage at the lacerating waste of life so carelessly dispensed and so casually snatched away. Yours is the sorrow of the sensitive soul that mourns for all living things that are handed death before they are given any kind of meaningful life.

"One could say that we are the lucky ones—the privileged few who are able to see outward and comprehend some of the wonders and beauties around us. Yes, we are lucky in that way, but we are also cursed with the ability to look ahead and grieve for the long silence and the coming darkness not only for ourselves but also for those we love and countless others we see whom we could not help.

"Is the beauty, the light, the love we experience enough to compensate for the suffering, the tears, and the grief—not only ours but also that of the numberless other living things that appear briefly and then so without notice are swept away? Sometimes we answer yes—sometimes we hesitate, struck mute. And always ... always ... dark-robed Grief is standing by, watching from the shadows.

"Still, we are the lucky ones comforted by the beauty and the love in our lives. But even if we cannot legitimately grieve for ourselves, how can we not grieve for all those countless others who are given no respite from the brutal side of existence? If it were not for the immensity of all that we do not yet know—if it were not for the hope that there is much more than we understand that might somehow justify all that we see—we might be tempted into despair. I am glad you did not go that way, Sandra—that your life force kept you on course for a life of helping others in their pain. I am happy that I was chosen to be your contact, and I am honored to know you."

With that and other similar conversations, Sandra's affection for Grace became love, and it became unshakeable.

Chapter 22

CHATS WITH GERRY AND WALTER

Gerry's Visitor contact identified himself as Walter, the same name as Gerry's grandfather who was a dear memory from his childhood. Walter obviously enjoyed Gerry's enthusiasm and deep interest in all things about the Visitor's technology. But out of necessity, Walter became adept at turning aside Gerry's persistent questions about points of knowledge that the Visitors deemed humans not yet ready to know. One time Gerry pressed Walter for details about the Ship's drive engine, something that obsessively fascinated him. How could it possibly work? What an incredible boon for humanity it would be to have such a technology.

Walter answered him this way—"Gerry, for now it is enough for you to know that a thing is possible. That alone is a huge advantage; sorting through the infinity of the impossible takes much longer."

Even though Gerry was often frustrated in his quest to understand some of the Visitors' secrets, there were enough moments of laughter and fun in their conversations to keep him happy.

One of Gerry's favorite topics was on the nature of physical reality—What is spacetime "made" of? How can it be visualized and interpreted? And how do the fields we have identified interact with spacetime at the most basic levels? Do the fields themselves comprise spacetime or at least space? Or are the fields properties of some yet higher dimension and only overlay or penetrate our spacetime from there? One day Gerry asked Walter, "What do you think about humans' ideas on spacetime down at the level of Planck length and Planck time? And what about the ideas that there might be additional dimensions besides the ones we experience?"

"Hmmm, more dimensions ... OK. Well, let's back up a little," Walter said. "I know that you read that entertaining book *Flatland* when you were young and that you have thought a lot about the possible existence of higher dimensions different than the four we are accustomed to in spacetime."

"How did you know I had read that book?" Gerry asked.

"Oh dear," Walter said. "I guess this calls for an apology. I have read all the zmails you have sent, and I recall that you discussed it with a friend of yours several times. I apologize if I have been too intrusive. I thought that anything available in your internet 'cloud' was available for study. I'm sure you know that we want to understand you as much as possible."

Gerry was nothing if not easy-going and forgiving. He replied, "Oh well, I should have guessed that, Walter; I know you're a good guy. I suppose I haven't said anything in a zmail I would be ashamed of—plenty of ridiculous things sure, but nothing too embarrassing. So never mind. Let's get back to higher dimensions."

Walter said, "Good! This is fun for me. Let's start with the imaginary Flatlanders who exist only in a two-dimensional plane universe. If some superbeing who exists in a three-dimensional universe were to look down on them and push a three-dimensional object into their two-dimensional world and then rotate the object sideways, what would be the effect? Let's say the object was a cone and its pointed end was pushed through first. Since Flatlanders distinguish each other by their two-dimensional outline that they see or feel by moving around it, they would be astonished to see this new circular individual suddenly appear out of nowhere and incomprehensibly continue to grow larger and larger. And then, as the superbeing begins to rotate the cone sideways, the Flatlanders would see the perfect circular individual begin to distort horribly into a changing series of ugly conic sections and at last stabilize as a completely different, but recognizable individual—an isosceles triangle, which to them is a soldier and possibly a dangerous one. The superbeing, of course, is quite naughty to disturb the poor Flatlanders in such a way."

Gerry: "Yes, I had a lot of fun with this book when I read it as a kid. At the time I was too young to appreciate its acerbic satire of Victorian England's class system. I read it again as an adult, and it was even more fun."

"Right," Walter said. "It's very good for reminding each of us how limited our point of view might be. Keep that in mind while we go back to spacetime. You might remember we mentioned in one of our earliest answers that we are bound into its four-dimensions just as you are. We are just as fascinated by it as you are."

"Sure—good," Gerry said.

"OK, so you've been taught that the most fundamental entities of the universe are fields, and fundamental particles can be thought of as excited vibrations in those fields—vibrations that appear to be stabilized somehow—and it's not known how—to persist through time and follow certain rules of interaction with fields and with other particles. You asked about the idea that our familiar spacetime might be at its most basic level a highly interconnected network at the Planck length scale, and that's as small as it goes—down at 1.6×10^{-35} meters. That scale is down at the level of the 'graininess' of the universe so to speak. That concept is basic in your science—that space is not continuous and infinitely divisible, but rather, its smallest fragments are Planck spaces no smaller than bounded by one Planck length.

"Also field strength values are not continuous and cannot have smaller divisions than a discrete value in each Planck space compared to the discrete value in an adjacent Planck space. You could ask, what is this network comprised of—is it a property of a field or is it only an abstract idea representing a field? Or rather, is the network perhaps something more primal than a field—something that has properties from which a field emerges? Does each kind of field have its own network or do all fields emerge from the same one? And apart from those questions, we can ask, is this network truly here all around and through us like an invisible, ghostly crystal lattice? Are we like a fish that swims though water unaware of what water is?"

Walter paused before continuing. "Well that's one view, but what about another? You've read about this one too. What if that network exists in a different dimension beyond our perceived 3D space, and our experienced spacetime emerges out of the interconnectedness of that network—somewhat analogous to a three-dimensional scene emerging from a holographic pattern of spots arrayed on a transparent film?"

"Whew, Walter!" Gerry said. "A lot of questions there. Yes, I've read about that last idea. It's been around for a while—it's fascinating. Of course the holographic analogy only helps a little. The idea you mentioned proposes that our actual, experienced reality is literally a projection from a pattern in some higher dimension."

Walter replied, "That's good—you've got that. We're heading there. You know how the field values in some areas of space can be entangled with the field values in other areas—the phenomenon of quantum entanglement?"

"Sure, of course!" Gerry said.

"And in empty space," Walter continued, "the degree of entanglement is directly related to the distance between regions of space and, therefore, relates a kind of geometry to quantum states. The closer two different regions of space are, the more entangled they are. And recall that entropy is directly related to entanglement. More than fifty years ago one of your physicists saw a relationship between entropy and geometry, and from that he was able to derive Einstein's equation for general relativity related to entropy—the beginning of a study called 'entropic gravity'."

Gerry said, "Yeah, that was a surprising insight. And about the same time, another group examined how to think of entanglement in terms of quantum degrees of freedom—an abstract way of considering entanglement that could be applied to things like properties of fields. If I remember right, they also were able to link that concept to a corresponding geometry and show that it also conformed to Einstein's equations for general relativity, at least where gravity is weak and spacetime is relatively flat. Hey Walter! This is exciting—getting on the road to linking gravity and quantum physics. Are you about to relent and hand me a big reveal after all?"

"Ho ho ho Gerry. Nice try," Walter said, "but I would never deprive you of the pleasure of discovering it on your own. We're just having a nice discussion here, and as you know, I need to keep it floating in the pool of your current science. We have far to go with many forks still in the road ahead, but I'm not above tossing out a little crumb here and there."

"Well yeah, absolutely!" Gerry replied. "And I'm counting on you to forget yourself one day and accidentally drop a really big one on me." They both had a good laugh over that.

Walter said, "So let's go back to thinking about dimensions. We know that each dimension of our 3D space is perpendicular to the other two. And if Flatlanders happened to live in any two dimensions of our three, they of course would not be aware of the third. But if they could 'look' in the right direction from any Planck space in their 2D plane, that third dimension would always be perpendicular to them no matter where they were. And even if they could sense it somehow, they can't enter it."

"Yup, I get that," Gerry answered.

"OK," Walter said. "Let's leave the poor Flatlanders behind and think of ourselves now in 3D space and just as frustrated with trying to visualize another higher dimension—say a fourth. We often talk of the fourth dimension as being time and think of it as joined with our three-dimensional space—hence the word *spacetime*. And, indeed, we often think of time as a kind of spatial dimension. After all, it distorts and flexes under the influence of gravity just as 3D space does. But it's also different. Think of yourself as a big particle—a mass of subatomic particles held together by the usual forces. So this big particle—namely you—can bounce around in any direction you please in 3D space. But you do not know how to do that in the time dimension. You can only drift helplessly along in one direction—forward into the future. The only flexibility you have is to slow your rate of travel through time by taking on enough relativistic energy as velocity through space or by entering a powerful gravity field. Why do you think that is?"

"Whoa, you put me on the spot there, Walter. Let me think ... Well OK, maybe because the time dimension cannot contain particles per se—maybe it contains other kinds of stuff ... OK, not stuff ... but ... well—hey wait! Maybe time is that higher dimension we were talking about earlier that contains the pattern—you know, like a holographic pattern—from which our 3D space emerges."

"Now you're getting into it, Gerry!" Walter said. "So then how does the change we experience as *Time* arise out of that view?"

"Well, I wonder ... that pattern might be a pattern of quantum entanglement at or between every Planck space in our 3D universe. Ah yes ... the current instantaneous time point of our reality could be imagined as the precise pattern of entanglement that *defines* our present instant of reality—all of the interconnected quantum states that exist at the present Planck instant. We could imagine that discrete pattern as having just resolved from a state of superposition one Planck time unit ahead of us to become our current instant of reality. Then in terms of quantum physics, there would be other patterns of entanglement, perhaps an infinite number of them, still there in superposition to the current instant. So stepping ahead one more Planck time unit to one of those alternate entanglement patterns in superposition would cause it to resolve and become our new current instant of reality. So our experience of the 'flow of Time' could be like the individual frames of a movie film passing one frame to the next to give the illusion of movement—stepping along one Planck time step to another—one entanglement pattern to the next in superposition. Hmmm ... I hadn't thought of it quite that way before."

"Well done, Gerry!" Walter said. "That's not bad for just off the cuff. You must have had your coffee this morning. But don't get the big head yet. There's a long way to go. The next question that will occur to you is—is there really an infinite number of entanglement patterns still in superposition and, if so, what determines which specific one is the next in queue to resolve into our current reality? And of course, what is the propulsive force—the drive that causes stepping forward from one Planck instant to another? We'll get back to that another time."

And so it went with many such chats between Gerry and Walter. On another day Gerry talked with Walter about a conversation Sandra had told him about with her contact a few days earlier when she learned that the Visitors envision the possibility of someday discarding their biological bodies entirely. He asked Walter what they see as their future in that all-machine state.

"That's right, Gerry. It's an idea we've thought about for a long while. Our motivation comes from our desire to penetrate particular deep secrets of the universe that have so far beaten us back. You've heard we have a very capable artificial intelligence system on our ship, but it's far short of those we have created in our home region. At home through our pod interfaces we have been able to achieve a kind of merging of ourselves and our AI networks. When we do that, we experience glimpses of a greatly enhanced level of consciousness, awareness, and cognition. And the words *greatly enhanced* do not adequately describe it—it's quite wonderful. It's as if we enter into a new and greatly expanded reality. But we have not yet mastered what we need to do to make it stable and safe. Our desire to overcome these problems is very strong. We think part of the problem lies in limitations of our biological brains—possibly the speed of signal processing they are limited to. So we are presently exploring that question. If we succeed, we hope it will open up vast new areas of understanding for us. We hope that it might even permit us to visualize and directly experience higher dimensions perhaps through entering into a kind of mathematical reality space that so far we have only sensed as glimpses."

Gerry was quiet for several moments and then said, "I'm overwhelmed. In one sense I'm so envious and want to experience that enhanced consciousness too, but in another sense I'm so happy for you—that you have the possibility to achieve it. It truly does sound wonderful."

Walter said, "Thank you, Gerry. Yes, you're right. We are both thinking creatures, and for us, understanding is the most addictive of drugs."

"That's a perfect summation, Walter!" Gerry said.

Then he said, "Would you be able to tell me about some of those deep secrets of the universe you are investigating?"

"Well let's see," Walter began. "Most of them wouldn't make any sense to you yet, but ... oh yes; I just thought of one that would be perfect to begin with. We were talking about it the other day. Do you remember when we were talking about the experience of the flow of time as being successive steps, one Planck instant per step, from one existing pattern of entanglement to another pattern still in superposition?"

"Yeah sure—I remember that," Gerry answered.

"Good. Let's consider a dynamic situation," Walter began, "say two billiard balls on a level billiard table. To start our scene, the cue stick strikes one ball sending it rolling to collide with the other ball. Recall we said that the degree of entanglement is greater for spatial regions closer together. So as the rolling ball approaches the stationary ball, the patterns of entanglement between them increase Planck step by Planck step. Each pattern in superposition that resolves into the next current instant of reality is the correct pattern to precisely specify that position along the trajectory of the ball. It includes both the translational and angular momentum and both the translational and angular kinetic energy of the ball. And, indeed, the pattern must include every feature like those plus every other feature of every fundamental particle that is a part of the ball. At last the balls collide and rebound resulting in the stationary ball now moving and the first ball suddenly moving in a different direction. The total original momentum and energy of the first rolling ball is now distributed between the two balls and precisely specified in the entanglement pattern of each. At the instant the balls touch they are as close as they can be, and the level of entanglement for their positions is at a high point. After that, the changes in the direction and momentum of the two balls would be seen in the new patterns of entanglement that are changing a Planck instant at a time for the two new trajectories."

"Hey, I might actually see where you're heading," Gerry said. "We know that billiard balls rebound predictably. But if this interpretation is right, the balls are not the primary thing—the entanglement patterns are. Why do the right patterns snap into place at each Planck step to create the predictable trajectories?"

"Excellent leap, Gerry!" Walter said. "That's good—we want to find out how the rules of changing patterns of entanglement work. Out of maybe an infinite number of patterns, why do just the right ones resolve into reality at each Planck instant for material objects that have predictable interactions? The dynamics of massive objects like billiard balls or planets are predictable, but for subatomic particles, not so much. We want to understand at a much greater depth the details of how this works. We want to know if there are hidden structures that result in necessary rules of interaction. We hope it will shed light on our concepts of cause and effect. It's a big issue for us."

"Wow ... I see what you mean," Gerry replied. "A deeper insight into that could change many things about how we understand physical reality."

Walter said, "Yes, there are many other things too about entanglement that we want to understand better. For example, think back to the Flatlanders. If the mischievous three-dimensional being thrust the two tips of a horseshoe-shaped object into Flatland, the Flatlanders would see two beings appear magically a distance apart. Let's say the object had properties that could be changed by touching it. Then if a Flatlander touched one of the new individuals and it changed its property, they would see that the other individual in a different location also changed instantaneously in the same way. It's a poor analogy to entanglement phenomena we see in our spacetime, but we wonder whether similar things are happening all around us but are hidden in another inaccessible dimension. It keeps us busy." More laughter.

"You know, Gerry, we haven't worked everything out yet either. We're still thinking about many of these problems and are still running experiments to find direction from new data. Our ship is not only a home for us on our voyage but is a laboratory workplace as well. We have many research instruments and facilities for studying different aspects of basic physics as well as the other sciences."

"But you've been working on these problems for thousands of years. If you haven't cracked them yet, what hope do we have?" Gerry asked.

"Don't be discouraged," Walter replied. "We have solved many puzzles, and you will too. Some of those solutions led to our development of the drive for our ship. Like us, you will be able to travel to other stars someday when you find the solutions you need. But over past centuries, when we found new solutions, we also found new questions. The new puzzles were just as tantalizing as the old ones. You will find it to be the same. I think both of us will always find a reason to keep searching to satisfy the hunger of our curiosity. You will need new ideas and new words to describe them. The mind can only understand what it is prepared for. The new idea and its new word must first be absorbed and understood as a step on a ladder to support you to a higher step of yet another new idea springing from it. Notions of symmetry, balance, and parsimony can be used as a guide for thinking, but thinking often must go beyond those. The relational aspects of reality will be important, and mathematics can provide guidance if your theoretical models are on the right track.

Only actual data though—tests of reality—can be the judge to rule on which idea to take seriously."

"Do you think there is no end to these kinds of questions?" Gerry asked.

"We'll likely never know the answer to that question," Walter said. "For me it helps to remind myself to be humble. I still feel that way when looking out into the vast mysterious voids between the stars. A thought occurred to me once that I can translate into something you probably experienced as a child. Imagine having captured an interesting insect and putting it in a glass jar to study it more closely. Imagine how that little creature might feel if it had enough consciousness to wonder what had happened to it. It might wonder why it can see the outside world but cannot move toward it—a wall that it can only feel and not see forces it to walk in circles. It would have no hope of understanding what glass is composed of or why glass is a solid but is still able to transmit light. It would not be able to understand what kind of thing put it into its transparent prison, and it would not understand why it had been put there. Our cognitive abilities are greater than the insect's, but in relation to the scale of the problem we seek to understand—perhaps we are just as diminutive as the insect. As odd as it might sound, the humility these thoughts engender does not discourage me. Somehow it helps me push on."

Gerry absorbed this in silence for a few moments and then found relief with, "Walter, ol' pal, you give a great pep talk." They both laughed.

Then Gerry thought to ask something he had wondered about since he first met Walter. He said, "Walter, would you mind if I asked how old you are?"

"No, I don't mind," Walter said. "I am getting on a bit as you might say, but I think I have a good while yet in the game. Let me think ... I need to translate my age into your years."

"Sorry, Walter, I forgot that you need to shift your time units into mine and your base-eighteen number into my base-ten number," Gerry said.

"No worries, Gerry; I have it now. So, I am now 39,873 of your years old."

Gerry's eyes widened a little, but he had expected as much. He said, "Congratulations, Walter! That definitely makes you my oldest friend."

They both laughed again.

Chapter 23

CHATS WITH ELLEN AND NIA

Ellen's Visitor contact introduced herself as Nia, the same name as Ellen's aunt who had lived nearby when Ellen was a child. Ellen's Aunt Nia had given her many happy hours of playtime and loving companionship, and she secretly imagined that this new Nia might be much like her dear aunt. She was not disappointed.

In the aftermath of Jim and Ellen's return from Australia and in the midst of the constant drama of the Visitors, Ellen struggled to focus on her work at the university. She did succeed in taking good care of her classes and her three graduate students, but she was having difficulty continuing the book she had begun writing before the trip. The book was about the history of concepts of human rights and how the circle of people regarded as like oneself had gradually widened over the centuries from family to tribe to nations and finally to include ideally all human beings. Other books on the topic already existed, but Ellen thought the subject was so important that it needed to be updated and reinterpreted in different ways to keep it alive and relevant. Inevitably, the subject of her book came up in her conversations with Nia. The Visitors were already famous for encouraging the very ideals that Ellen wanted to write about.

In Ellen's third conversation with Nia, she said, "In your earlier messages to us last January, you spoke of your harmonious society and its ideals of fairness. I have been trying to write a book on that subject and wonder if you would be willing to talk with me about how your people have worked to achieve such harmony. Would you mind if I asked you some questions about it?"

Nia answered, "I wouldn't mind at all! That subject is dear to all of my people and one I enjoy talking about. What would you like to ask first?"

"Wonderful!" Ellen said. "I have been wondering whether your people evolved somehow to be naturally more fair and empathetic than humans or whether in your history you might once have been more like humans were and still are. For example, I wonder if your people in the far past ever engaged in slavery? That is a practice that was nearly universal in early human societies and persisted into more recent times. And even now it still is practiced in an undercover way in some cultures."

"You have started with a good question, Ellen," Nia answered. "Yes, we once were much like you. In a time long before the Old Ones and their Great Understanding, we had evolved into an intelligent species that struggled to dominate its environment to its own advantage—a pattern that seems to be the norm. In those times there were power struggles between different groups and wars between different nations. There was much suffering and waste of lives that persisted for many thousands of years. During those times, slavery was common. In the literature of that period, the winners of a war justified enslaving the survivors of the losing side by saying that they were enemies who had made war on them and, therefore, deserved to become slaves. The question of which side started the war or for what reasons was avoided. It was simply the way of the world, and the winners always proclaimed that their side was just. Eventually any group that was too weak to defend itself against a more powerful group was deemed to deserve being enslaved on that basis alone. It became a saying—*the weak are born to serve the strong.*"

"Did some of the slave holders have second thoughts about the justice of what they did?" Ellen asked. "Did no one see it as a form of injustice?"

"That question was not recorded in the earliest writings. Much later there were a few who were more empathetic and more sensitive to the plight of slaves. But the vast majority of those in power fought vigorously to maintain slavery and were happy with their own privilege and status. A long period of argument, persuasion, and appeals to justice was needed to sway more minds away from slavery."

"What kinds of arguments were used to change minds?" Ellen asked.

"The most important thing to do first," Nia answered, "was to persuade slave owners to recognize that slaves were persons like the masters. All were members of the same people. For example, it was not unusual for children to be produced through liaisons between a slave master and a slave. Still, it is no surprise that slave masters were peerless at justifying themselves by inventing reasons why they were superior in one way or another to the slaves. But the reasons never stood up to rigorous logic. Unfortunately, logic did not seem to matter for a long time. People who were advantaged by slavery managed to find new forms of denial to justify their position."

"That story is very familiar from Earth's history too," Ellen said.

Nia went on. "Another important argument was to remind supporters of slavery about how slaves came to be slaves. Were they guilty of some crime so terrible that they deserved to be punished so severely? No, they were victims of forced capture or of an unlucky birth to someone who was already a slave. As nations grew and gained power, it was no longer common that one nation attacked another in war and then lost that war. It had become much more common for a powerful nation to attack a weaker one, and then take what it wanted including those survivors they captured. The captured people were innocent of offense; they simply did not have the ability to defeat the attackers. And apart from nations waging wars of conquest, there was no shortage of smaller groups of mercenaries and pirates who staged lightning raids in peaceful territories and kidnapped large numbers of innocent people while murdering many others in the course of the raid. The raiders sold those kidnapped victims to others who then sold them into the ownership of slaveholders. Even in those early times, it was against the law to knowingly purchase stolen goods. But slaveholders refused to acknowledge that those poor people standing for sale in the slave market were only there because they had been stolen from their homes and their communities. And beyond that, what justification could there be for enslaving children born to those who were already slaves? What offense had the children committed to deserve such a fate? Only a few people who saw the slaves as individuals like themselves were able to see the logic of these arguments." Nia said.

"Yes, that is identical to practices that were common on Earth," Ellen said. "I remember reading about the Ottoman Empire when, in the fifteenth and sixteenth centuries, soldiers were regularly sent into Eastern Europe and the fertile plains of an area known today as Ukraine to 'harvest' people to be sold into slavery. The merchants who bought and sold them used to joke—'Can there be any people left there?' The soldiers actually practiced wild game management strategies by rotating their raiding sectors to give recently 'harvested' areas time to recover their populations.

"And in the early history of my present country, records show that twelve-and-a-half million people were captured and stolen from their communities in Africa to be forced into slavery in the Americas. It adds to the horror that nearly two million of those people did not survive the trip across the Atlantic Ocean but died from the terrible conditions of their passage trapped naked in chains deep in the hulls of wooden ships.

"In your history, was an argument ever found that seemed more effective in changing the minds of supporters of slavery?" Ellen asked.

"Yes, there was one that did begin to change minds," Nia answered. "You could call it 'the argument of role reversal.' It underlies what you call the Golden Rule, and in many respects it underlies all of morality. For this case, it asked slave holders whether they would be content if raiders came into their communities and captured them and their families and sold them into slavery in another country. Would they consider that their fate was just—that they should remain slaves for the rest of their lives and also that their children and grandchildren too should be slaves all their lives and so on?

"Only people willing to listen took the question seriously. But it was a question that could only be ignored, not one that could be dismissed with logic. Over time it did have an impact. But still, the central issue always was to convince everyone that slaves and slaveholders were the same kind of being—members of the same species, each able to suffer like the other and each as deserving of freedom and justice as the other."

"In Earth's history we have had problems based on people's differing skin color. The visible skin color difference between some groups made it easy for some people to suppose that people of a different color were not the same as themselves and, therefore, did not merit consideration as equals. Did your people ever need to deal with a situation like that?" Ellen asked.

"Not a situation of color," Nia said, "but we did have one group that was physically larger on average and more robust than others. For part of our history the larger form claimed to be superior beings and for a time made war on that basis. But eventually it became clear that the difference was only a modest local gene pool difference caused by a founder effect in the past combined with a geographically isolated environment that selected for greater body size. Large and smaller types could mate and produce perfectly healthy children of intermediate sizes. As improving technology allowed our populations to move more freely around our home planet and interbreeding took place more frequently, the difference disappeared. Later genetic analysis proved unequivocally that we all were of the same species. Certainly that should always have been obvious to anyone willing to look with an open mind."

"Yes, that is exactly the case on Earth with our color differences," Ellen said.

Nia responded, "That color issue by the way has been a puzzle to us. It is so obvious that the higher melanin pigmentation for some groups is very important for protection from your star's fierce radiation. In equatorial regions of your planet the radiation is more intense and selected for gene pool differences in favor of more melanin in the skin. And yet you don't seem to be concerned about eye color or hair color perhaps because those differences still occur commonly within various groups."

Ellen said, "I can see why you would be puzzled—it puzzles us too. It seems clear that it's a case of people with greater power wanting to protect their privileged position by keeping another group in a subordinate, weaker position. For example, in the American South and even some places in the North in the eighteenth and first half of the nineteenth century, it was against the law to teach a slave how to read and write. And prejudiced attitudes against people with darker skin have persisted even into the present. Only thirty years ago there was a series of massive protests against systematic racial bias across the US. Although change for the better has been painfully slow, at least there has been some improvement. Sixty years ago it was worse than thirty years ago and today it's better than thirty years ago. There still remains much to make right, but thankfully there have been many people for a long while who do not understand why skin color should matter at all—skin color should be the way hair color and eye color do not matter."

"Yes, that's the way we think too," Nia agreed.

Ellen continued. "That whole issue seemed ridiculous to my White father. When he was working for a time in Africa where he met my mother, he didn't care about the color of her skin. He fell hopelessly in love with her, and she with him. They have had a happy life together. I have been lucky too in finding opportunities and not being held back. And my husband loves me for who I am. It wouldn't matter to him whether I was blue, green, or brown."

Ellen found these conversations helpful in bringing her mind into focus for her book. One day she asked Nia, "How did it happen that your people finally came to accept that they really were all one people and all deserved equal treatment? Was there some pivotal event or some particularly powerful argument that helped everyone to understand?"

"The answer is yes and no," Nia said. "There was a special period in our history when new ideas of tolerance, empathy, and fairness began to appear more frequently, but it did not happen overnight. It took time and persuasion for the ideas to incubate and be accepted. It was still in the early times of the Old Ones. Eventually the ideas matured and developed into a norm, and we came to call that period 'The Great Understanding.' After this we were finally able to end the wars that still happened. It was also a time of great technological advancement and prosperity that promoted acceptance of the new ideas by making better communication possible and by making the necessities of life and material benefits available for all."

"That must be a fascinating period of your history! Would you be willing to tell me more about it? Would you mind if I included the story in my book? It might be a helpful guide for Earth."

Nia said, "My answer is an enthusiastic YES to both questions!"

And thus began Ellen's determination to write the story of the Visitors' "Great Understanding."

Chapter 24

CHATS WITH JIM AND GEORGE

Jim also became close to his Visitor. In Jim's first call, the Visitor introduced himself as George, the same name as Jim's dearly remembered father. But that name was even more appropriate to Jim because he had another favorite George as well—George Price, a famous contributor to his field of the evolution of altruism. Jim's interest in the evolution of altruism developed partly because that issue is such a complex and intriguing problem and also because of Jim's unusually great capacity for empathy. Jim's even, good humored personality on display to others had always hidden a core of anguish at the suffering and injustices of the world. Jim liked to think that science could demonstrate support for moral insights and, therefore, strengthen their impact. It did not take him long to raise this topic with his contact George. As was the case with all the Visitor contacts, George was very knowledgeable in Jim's special field. Jim and George were soon immersed in discussions of evolutionary theory.

One day Jim said, "George, I know it's been many thousands of years since your people faced natural selection in the environment of your original home planet, but did the Old Ones leave records of their studies of how evolutionary development might have happened in their earlier history?"

"Yes, indeed, they did," George answered. "Well before the Old Ones left the original home planet to live in space habitats, they had developed a highly technical civilization that existed for thousands of your years despite the intrusion of periodic wars. Their science from that period documented their studies of life's history on our planet, and they developed theoretical frameworks for their findings just as you are doing. It is one of our greatest treasures that virtually all of their old knowledge has been preserved and is still with us today."

"That's wonderful," Jim said. "You're so fortunate to have such a complete history. I know you have studied Earth's science on this subject, and I've been wondering whether broad parallels exist between your history and what we humans have found of Earth's history."

"There are certainly broad parallels," George answered. "And from your question, I'm sure this is not a surprise to you. What we have found here and on other life-bearing planets is that the same general patterns of survival and competition prevail wherever life forms can reproduce vigorously. In natural environments, life forms face contingency at every level. What you see of biological nature on its surface are the survivors playing their parts in the drama, but it is only a snapshot in time. It's true that a later snapshot might look much the same, but the individual actors will have been removed and replaced with their offspring. In the larger scale, it is always this way except when the stage is wiped clean for a while by major natural disasters."

Jim said, "Yes, that's the overview we see here too. What really intrigues me though is how altruistic behaviors are observed in many species of social animals here—a behavior where one individual takes on a greater risk or sacrifice to itself with an action that benefits others in the larger group. Examples of cooperation and helping behavior among animals are everywhere. It's fascinating because it seems inconsistent with each individual struggling to maximize its own chances to survive and reproduce while competing with others doing the same in the face of limited resources. One expects the selfish members of the group to outcompete the self-sacrificing member leading to the extinction of that altruistic trait in the group. For well over a hundred years here on Earth, geneticists have tried to explain this phenomenon by working with mathematical models. A lot of progress has been made, but even after all this time, different interpretations provoke controversy and disagreement."

"Indeed!" George replied. "The scientists on my planet who first studied this question had the same experience. I looked through your records and saw that Earth developed good working theories for the evolution of eusocial insects like bees, ants, and termites. Genes giving rise to self-sacrificing behavior are preserved and increased in the population because they benefit the group as a whole and the insects in their groups are so closely related to each other. The altruistic traits are strong too; specialized soldier ants and soldier termites will fight to the death to protect the colony."

"Right!" said Jim. "With insects, we think of them as reacting automatically to a stimulus like an intruder with no choice to disobey the defensive instinct. But I have wondered whether animals with more complex brains could have more flexible responses that add another layer of complexity to consider. I've read many papers using various models under different theoretical approaches to interpret how selection for altruism might occur—conceptual approaches like inclusive fitness, reciprocal altruism, multilevel selection, and others. I realize that correlation models look at the expression of a measurable trait that can appear in and disappear from a population, but I have wondered if an expanded view of the trait might be helpful."

"How do you mean?" George asked.

"Well," Jim said, "if, for instance, competition from selfish individuals drives the new altruistic trait into extinction, how long would it be before that trait might appear again in the population? The models usually include a factor for how often the trait would randomly appear, but it seems to me that the assumed frequency of reappearance is too low. And that issue might become relevant to the model if other factors available to animals with more sophisticated brains are not ignored? I wonder if an oversimplification has overlooked a deeper puzzle."

"I've read the older papers. The modeling is valid for the limited conditions the equations describe. Are you suggesting that considering only the outward expression of the trait is overlooking something important?" George asked.

"Yes, it seems that way to me," Jim said. "Think of the granularity of the real world; it's complex and messy. So we define a cleaner situation in mathematical models and ask it to stand in for reality while expecting that the messiness of myriad real world interactions will blend over and mostly cancel each other out in the background while leaving your chosen variables intact. One would like to treat these problems like say the clean derivation of the ideal gas law from the kinetic theory of gases. That's what we have mathematical tools to do."

"Tell me a little more Jim," George said. "This sounds interesting."

"OK, I've been thinking that I might need a different model because I've been thinking about a different way an altruistic trait could arise in a population of more complex animals. Let me expand that first with an example I have in mind. I read a paper once by a primatologist who was describing some of the differences between humans and chimpanzees, our most closely related primate. She said that you would not be able to put three hundred chimpanzees crowded together on an airplane and expect them to sit peacefully and in reasonable harmony for a multihour journey. Before long, probably before the plane left the gate, squabbles would have broken out, and in no time a full-blown melee would be in progress. In the eventual aftermath, the cleaning crew would need to collect bitten-off fingers, pulled-out hair, clean blood off seats, and possibly remove a corpse.

"By contrast and as violent as humans certainly can be, we are generally more cooperative and less aggressive than the chimps. How did natural selection shape the two primate lineages that split off from our last common ancestor with the chimpanzees?" Jim asked. "Somehow a difference in social cooperation and aggression levels as well as a difference in cognitive ability developed between the human line and the chimp line. We can only speculate on how natural selection brought that about. The main point though is that the last common ancestor would already have been a complex social animal with a suite of genetically inherited social behaviors that it passed on to both the chimpanzee line and the human line."

"Yes, I would agree with that," George said. "After the split, the human and chimpanzee branches evolved independently and modified what was inherited from the last common ancestor."

"Right," Jim said. "And I would expect these social behaviors to be inherited as polygenic traits—each trait encoded by at least several or more likely many genes. That's significant because the genes encoding each particular trait would themselves have variant forms in the gene pool—alleles—that in different combinations in a single individual would express that trait to different degrees. So, like height or skin color, each of these behavioral traits would be expressed differently in individuals depending on how the genetic recombination fell into place for each new conception. In the population, for instance, if the trait was 'quickness to anger,' one individual might be very easy-going, another could rouse to anger if provoked enough and another could be hot-tempered and on a hair-trigger. Other individuals would fall somewhere in that range depending on their gene combination. These inherited social behaviors are better described as emotional tendencies rather than specific actions that cannot be disobeyed like those in a processionary caterpillar or a soldier termite. So why am I going through all this background?"

"I was wondering that too, Jim," George answered. That brought a laugh from both. "But seriously, I think I know where you are headed. How about spelling it out for me?"

"Sure!" Jim said. "Just drawing a breath. First let me lay out a little more context. I should identify some of the behavioral traits I'm thinking of. They would be emotional tendencies that would be triggered by environmental stimuli. Examples are quickness to anger that I've mentioned—also nurturing behavior, aggressive behavior, curiosity, empathy, indignation at unfairness, fearfulness, boldness, helpfulness, selfishness or, its counterpart, generosity. There are more examples of course, and I'm not suggesting that these few are rigorously defined. They may prove to be comprised of deeper behavioral elements, but I'll focus on their outward expressions for now.

"I'm talking about these genetic traits because I see them as a way for an altruistic trait to arise in and be maintained in a population. It's a different interpretation from the usual notion say of just a new mutation in a particular gene that can be lost from the population if the carrier does not reproduce adequately.

"Here's how I mean that. What if one individual happened to inherit a low level of quickness to anger, a low level of selfishness, a high level of empathy, and a high level of nurturing behavior? Well, from the outward expression of that blend, that individual could be seen as having an altruistic behavioral trait. And if that individual's tendency toward helpfulness, generosity, and self-sacrifice disadvantaged it so severely in the struggle for existence with other selfish individuals that it died without reproducing, the tragedy of that individual's loss would not mean that the altruistic trait was lost from the group's gene pool. The various genes coding for the various polygenic traits are still in the pool carried in the other individuals but just in different combinations. They can always be reshuffled randomly in future conceptions to produce a new altruistic individual from time to time.

"And in the meantime, the group as a whole may have benefited from the life of that self-sacrificing individual. That benefit could translate into greater reproductive success of the group. After repeated iterations of this same story over time, the group could begin to see the value of such altruistic individuals. In the early human line, brains were increasing in size and sophistication. That would promote better recognition and memory of helpful individuals. And that could bring forward more strongly another trait into play—reciprocal kindness.

Reciprocal kindness is a behavioral trait that is noted even in some lower animals.

"My point is that by considering an altruistic trait as emerging from recombination of other more basic behavioral traits, it is more likely that altruism could emerge, reemerge, persist, and have the opportunity to expand through other social forces. It means that it need not be thought of as a discrete trait that can appear only rarely and then disappear completely through competition from other nonaltruistic individuals. In time with more sophisticated brains, improved recognition of the benefits of cooperation could lead to a reproductive benefit for those exhibiting cooperative behavior. Then the genes underlying that behavior could increase in frequency in the gene pool, and the cooperative behavior would become more common—eventually even become a species trait."

"I like this line of thinking," George said. "I see now why you suspect that the older models might not have considered enough factors."

"Yes, lately I've been interested in more recent modeling efforts," Jim answered. "One group aimed to create a more general mathematical approach, but in doing so they also made it possible to address more of the particularities of real-world situations like the gene-frequency distributions of a population's gene pool, the population's reproductive mode and the spatial structure of the habitat where the population exists. I've played around with them for a little while now and have come up with some probability distributions. Then for fun I ran a series of Monte Carlo Simulations on those distributions and had some even more interesting results. But I'm not finished yet, and nothing is ready to publish."

"I would love to see some of the detail you've come up with," George said.

"Sure, George—can do. I'll try to organize it for you into some kind of sensible order. I need to do that anyway."

George said, "My dear friend, as you know I won't just hand out easy answers to you; I know you are better off thinking your way to them yourself. But I believe I will do no harm by saying that you are on a good track."

"Wow, thank you, George! That's a big help. I'll keep playing with it. I've been fascinated for a long time by the common emotional responses and cognitive biases that all humans seem to have. It makes sense to me that these features would interact with natural selection to shape what we think of as moral norms. The development of language that came to us later became the next great critical tool for creating complex cultures. That's where the plot really thickens.

"The thing I always try to remember though is that we still do not understand just how a suite of genes orchestrates a complex behavioral trait. We still have a lot to do."

"Don't get discouraged, Jim," George said. "These are grand notions that you love to play with. And someday they might lead you to a grand epiphany."

❖

So it went with the four friends. Each had found a dear new friend who also happened to be an expert in each of their professional fields. But all four had found much more. Each felt a deep and warm attachment and trust with their Visitor friend. When they stopped occasionally to think of it, they dreaded the day when they would need to say goodbye.

Chapter 25

CONSPIRACY

Washington, DC, Thursday, 18 July, 8:00 a.m.

By the time the third month had passed since the Visitors' Arrival, it was accurate to say that the Visitors' conversations had changed the human population significantly. No other means of mass communication in modern history had been able to achieve such an impact in such a short span of time. Although established cultural traditions in Earth's many nations persisted, most people recognized that their lives would never be quite the same. Many whose thinking had been limited to parochial interests had now begun to expand to a broader view of possibilities and found themselves dreaming about what life could be like in ways they had not imagined before. Historians who liked to write about major shifts in human history were already talking about 2052 as a tectonic year that would herald the beginning of a new age. Terms like a *New Renaissance* and a *New Enlightenment* were being casually tossed around in the media. Some even borrowed from the Visitors by calling the change Earth's *Great Understanding*. It was a heady time.

Earth's democracies found themselves dealing with changing attitudes among voters who were influenced by the Visitors. Progressive parties were having a field day promoting the messages of empathy, harmony and fairness that were commonly heard from the Visitors. Hardcore conservative parties countered with efforts to rein voters into their traditional concepts with strategies that emphasized the nonhuman aspects of the Visitors and attempts to separate humans from "them."

The central point for all though was an almost unconscious message that was not articulated but was always there as an undercurrent for all to sense every day and hear with every phone call from above. It was one clear thing—the stark, obvious power of the Visitors. It served as a focal point of fascination for nearly all humans. A gleaming ship overhead looking as big as the Moon—a ship representing a civilization incomprehensibly ancient by Earth standards and inconceivably immense—a civilization far far advanced beyond Earth's best science and technology—it was like a flame to a moth. Not every human was vulnerable, but for the great majority, the allure could not be ignored.

Moving into the fourth month of the Visitors overhead, President Kaitland and her administration were dealing with these changing attitudes and also with some recent intelligence reports that foreshadowed possible trouble ahead. The small disaffected minority that was feeling threatened by the popularity of the Visitors had been watching with growing anger and were plotting more countermoves. It was already known that a small number of well-funded international players were involved in the legal media campaigns against the Visitors. But some new information had just come to the attention of the CIA that these players might also be planning strategies outside the law.

On Thursday, the eighteenth of July at President Kaitland's regular 0800 meeting, the new intelligence came up. Dr. Jonathan Schneider, with his role in antiterrorism, had become involved with the reports and had been asked to report to the meeting. President Kaitland had great respect for Jonathan Schneider and was glad to see that he was already engaged with the issue. She began by saying, "Since the extreme tension we all experienced anticipating the Visitors' arrival, we've been able to relax a little and find hope that this recent peaceful period might remain with us. Over the last three months, the Visitors and their daily conversations with Earth's people have brought about a degree of peace and calm that we haven't seen before. So it is not happy news now that our intelligence agencies have received disturbing reports of a possible conspiracy that might result in extreme violence possibly to be blamed on the Visitors. Dr. Schneider, I believe you have the latest information to share with us this morning. The floor is yours."

"Thank you, Madam President," Dr. Schneider said. "Yes, these latest reports are most unwelcome. Snippets of information have been filtering in for the last week, but until yesterday nothing was very alarming. I know everyone here is familiar with the anti-Visitor media campaigns of the last two months. Earlier investigations had been initiated into the source of these campaigns as a watch-and-wait exercise. Although they were lawful as such, there was concern they could trigger violence. One particularly worrying group had transformed itself about the time the Visitors reached Earth. It emerged as a consolidation of reactionary groups mainly from America and Europe under a new shibboleth of 'Humanity First.' Early investigations had great difficulty discovering who was funding the campaigns. We did eventually find that media payments came from what proved to be dummy companies of at least two and sometimes three layers deep. Recent probes have penetrated behind the false barriers and identified four sources of funds—two from individuals in Russia and two from individuals in the Middle East. Investigators believe that these major players are manipulating for their own ends the passions of some of the other ideological groups. Our agents are confident that more are involved, but there is still work to do to identify them."

President Kaitland interrupted to ask, "Has anything been confirmed as to what kind of disruptive acts they are planning?"

"No, Madam President, nothing confirmed yet," Schneider replied. "But one local informant in Riyadh just passed on to us a report of an overheard conversation with one of the suspects. There was a reference to some kind of planned attack on the Visitors' ship. This report seems hard to believe though. At least it seems unlikely that anyone would seriously be planning such a thing. We do have enough information though to persuade us that a conspiracy for some type of violence is real, and the agencies are giving the investigation their highest priority."

General Beckworth had been listening intently and asked, "Dr. Schneider, has the investigation been limited to within US agencies or have US agencies already begun cooperative investigations with foreign intelligence groups?"

"I believe it is still sequestered within our own agencies," he said. "Why do you ask?"

Beckworth replied, "Well, like you I find it difficult to believe that any nongovernment entity would seriously plan an attack on the Visitors' ship. How could they do it? Even our own Earth Defense Shield might not be adequate against the Visitors. But it has occurred to me that a corrupt cabal might be able to manipulate a fanatical group to stage some sort of infiltration into one of the command chains for the EDS and possibly into one of the control rooms for missile launch. The Russian missiles have the most powerful warheads by a significant margin and would be desirable as a first choice for them. Their command posts are also the most closely located to where the known conspirators are living and to some of Earth's trouble spots where extremist attackers have been active. It's very far-fetched I know, but it's the sort of thing that sometimes keeps me awake at night. I'm picturing some fanatics ready to martyr their lives in order to take over a control room and force a launch of some of the EDS missiles against the Visitors' ship. I can't think of another scenario that could deliver a serious attack on it. I am aware that our Russian EDS partners practice a high level of security, and I wonder whether they have caught wind of any of this yet and whether they have any scope to increase security further. It would not be a bad idea for us as well to look for ways to increase our own security."

President Kaitland by then was looking alarmed. She said, "Yes, that is a nightmare picture to keep one awake at night, General Beckworth.

You were central to the development of the rules of engagement and the security measures to prevent accidental or any other unauthorized missile launch. In the scenario you just described, could that actually be possible? I mean, I understood that secret encrypted launch codes and at least two-person authentication and possibly more would be required for launch."

"Yes, Madam President," he replied. "All of those security steps are in place with even a couple of new protections. But the entire philosophy behind the development of the Earth Defense Shield was that it might be the only thing preventing the extinction of the human species. Critically, we all understood that it might need to be activated within seconds of receiving a command decision. Everything about how such a decision could be made was filled with uncertainty as we discussed at the time, and the consensus was that we could not afford an additional period of review and reconsideration in the launch control room. Every step in the process between receiving a final command and the actual launch would be done with the greatest urgency. The people involved would be thinking that they were acting to save humanity. So although safeguards were put in place, they could not be as redundant or as deliberative as we might normally prefer.

"I think that with enough money to corrupt a handful of key individuals and with a force of people willing to die in the attempt, it might be possible to force an unauthorized launch. With months of planning and preparation to infiltrate a control room, well ... " General Beckworth's voice trailed off into a sigh, and he seemed reluctant to complete the sentence.

President Kaitland said, "This suddenly is much more serious than it appeared yesterday when I saw the preliminary report. Imagine the aftermath of any scenario like this. What might be the ultimate cost to Earth of such an attack if any of the Visitors were harmed or worse? And what if such an attack caused the Visitors to counterattack us? Even if this scenario is far-fetched, we must consider it an existential threat. I want these investigations to continue with the highest possible priority. We must think through when and how we should alert our Russian and Chinese EDS partners, and we should rethink our own security standards and work out how to revalidate all involved personnel. And these are only the first steps. I want everyone to take four hours to review and analyze all of our evidence and come up with possible countermeasures. We will meet here again in four hours to develop a workable plan." She noticed General Beckworth seeking her attention. "Yes, General?" she said.

"There is one more question that I haven't heard raised yet. This might sound strange coming from me with my past suspicions of the Visitors. But if our investigations provide any more substance to this conspiracy, the question that comes to my mind is, shouldn't we also alert the Visitors?"

"Yes, indeed—that is the question that comes to mind," the president replied. She dismissed them and went to her private study.

Four hours later President Kaitland reconvened her meeting and called first on Dr. Schneider.

He began, "During the last several hours I have pushed every source we have for anything new. Since this morning, two new points have emerged. First, the two Russian individuals have considerable past relationships with people in the Middle East linked with rumors of paid assassinations. Second, all four of the suspects identified so far have vanished within the last twenty-four hours. We knew where they were the day before, and they seemed to be behaving according to past patterns. Then sometime yesterday they were gone. None of their passports have shown up as crossing any borders, and there is no credit card activity either. They might have assumed new identities with fake passports. This sudden disappearance is alarming because it suggests that some action might be imminent."

"I agree—that certainly is alarming," President Kaitland said. "Have our agents checked whether any associates of these four have vanished as well?"

"They were already in the process of finding that information when I talked with them a little while ago," Dr. Schneider said. "Within a few more hours they will have chased that down. If other of their associates have disappeared, that greatly raises the level of alarm."

General Beckworth joined with, "We need to find information about how old this plot is. It's reasonable to guess that it did not hatch until after the Visitors arrived and started making phone calls. That is what seems to have set off the conspirators. I mention that because in the initial creation of the EDS, only the most capable and thoroughly vetted personnel were chosen for any of the positions in the launch control rooms. We must check whether there have been any changes whatsoever in the originally selected personnel since the Arrival. We should probably check all the way back to the original activation date for the EDS in case the plot precedes the Arrival. And just as important, we should do the same for anyone who in any way might have access to be near the control rooms. We have three equivalent control rooms all in highly secured and widely separated locations, but there are service personnel, delivery people, military guards, and so forth who have duties near them. There are security procedures in place, but we need to consider that an attack could come through any avenue. Any recent change of personnel in those associated groups is a red flag."

"Yes, I agree. Thank you General Beckworth," the president said. "We must begin those checks as soon as possible. Several of you here are in the command chain for that. We'll set up the orders to begin the checking before we conclude this meeting. Dr. Schneider, you have something more?"

"I only wanted to add that I have arranged to be notified here in this meeting about any new information related to the disappearance of any of the four conspirators' associates. That news could come at any time," Dr. Schneider said.

"Good," President Kaitland said. "If we learn that others of this group have also disappeared, it's a signal for us and also signifies that the conspirators are a little sloppy. Let's hope that such a lack of attention to detail carries through to some of their other planning. We might soon find ourselves needing to make very consequential decisions very quickly.

"I think this is a good time to bring back the issue General Beckworth left us with this morning—namely, should we warn the Visitors about a possibility we have not yet been able to confirm. From what our intelligence sources have told us, the conspirators have apparently avoided committing anything to electronic communication. Everything specific to the plot that we know has come from local informants who overheard fragments of conversations. Therefore it is quite possible that the Visitors know nothing of this danger. I would like to hear discussion now as to the best strategic decision as well as an ethical decision. General Beckworth, you first thought to bring this up—would you like to weigh in first?"

"Thank you, Madam President," General Beckworth responded. "When I first thought of this question, it did not seem simple to me because of my earlier suspicion of the Visitors and the risk they posed. Despite everything they said about themselves during the first four months of this year, I felt it was my duty to continue thinking of them as possible deceivers and possible enemies. When they arrived with their astonishing ship and with claimed numbers onboard that could overwhelm us, it seemed that my worst fears were confirmed. But in the last three months we have seen nothing sinister. Everything we have seen has been just the opposite. There were no attacks, no demands, nothing to give us pause except perhaps their deployment of the seven relay satellites. But they used those satellites to reach out to us with nothing but helpful friendliness.

"Perhaps I could remain suspicious about the numbers they said were onboard. I could even entertain the notion that the voices we hear from the Ship are all generated by a powerful artificial intelligence. And even though I interrogated my own Visitor contact intensely for weeks and found nothing that seemed false, how could I not imagine that an immense and advanced AI system would easily be able to fool me? These are among the reasons that I should still be cautious. And from a different angle, if we tell them of a possible attack from rogue humans, are we exposing ourselves as a dangerous unstable species? And if we do not warn them, are we revealing ourselves to be treacherous and uncaring?

"This is not an easy one for us. With all that said, I suppose this is the moment that I should reveal that I am among the billions of people on Earth whom the Visitors have won over as friends. Despite all the cautions that might still exist, I would never fail to warn my friends of a serious danger if I knew of it. Even though we will do everything we can to stop the conspirators, how can we not also alert the Visitors so they can take steps to defend themselves or else quickly move beyond the reach of any of our missiles?"

President Kaitland had marveled at the transformation she had seen in General Beckworth recently and marveled even more at what she just heard him say. She had always been impressed with his competence and dedication to his job, but now she was seeing a more human and complete side to him and was moved by what he said. She answered, "Thank you, General Beckworth. I see that you have thought this question through carefully. I will hold back voicing my own thoughts a little longer until more have a chance to speak." She noticed Jonathan Schneider signaling. "Dr. Schneider, please go ahead."

"Thank you, Madam President. I'm impressed with what General Beckworth just outlined for us. I think he expressed well the conflicted thoughts many of us have as well as a sense of what feels like the right thing to do. A few months ago we were reminded of some events from our own history that are similar to our situation now. Since then I have read the historical sources describing the history of the Caribbean Islands following the first visit by Columbus and afterward by Spanish colonists. The native people of those islands would not have been able in the first few months to correctly interpret the intentions of the European arrivals. They certainly could not know the long-term fate of what would happen. The terrible fate that befell them in subsequent years at the hands of the Europeans is the story of most such contacts then and later in other parts of the world. We hope that we ourselves are better than that now. But we still cannot be certain whether the Visitors might turn out to be like those old Europeans or whether they are as they say they are. All of us here though have experienced their kindness over many hours of conversation—enough experience that would give us confidence in a human friend. I agree that if this conspiracy continues to look real, we must alert them and give them an opportunity to join with us in their defense. At least we do know with certainty that the rogue conspirators are not our friends."

President Kaitland thanked Jonathan Schneider and continued listening to the others. All of them had experienced many conversations with their Visitor contacts and had learned much more about them. She heard the same sentiments repeated again and again. When all had spoken, she said, "I'm glad it's unanimous then. You have expressed the same views I had arrived at myself. Thank you again, General Beckworth, for bringing it up.

"I have been wondering whether we should make this decision among ourselves or whether we should involve others—perhaps congressional leaders. But I believe this is an issue akin to a wartime emergency. We need to do what we can to save lives and do it quickly. We will begin with the intense checking of all the personnel recommended by General Beckworth. Then within the next twenty-four hours, if we receive further confirmation that—"

Just then she was interrupted by an aide from Dr. Schneider's group being ushered into the Situation Room. The aide walked to Dr. Schneider and handed him a piece of paper. Dr. Schneider read it quickly and then handed it across the table to the president.

President Kaitland read it and said, "I was about to say that if we received any other confirmation of the plot, we would need to alert our Russian and Chinese partners straight away. The intelligence report just handed to me tells us that at least seven other known associates of the four identified conspirators have also vanished in the last twelve hours. There may be more, and they are still checking on others.

"We have much to do now," the president said looking around the room.

One of the State Department representatives asked, "How do we go about warning the Visitors? They have never given us any detail of their government contact structure. Shortly after the Arrival they did identify a special frequency to use for government communications with them, but they seemed to suggest that it was more for ceremonial functions that only we thought were needed. There are twice as many of them on board as the entire population of Earth, and we don't even know if they have a captain of the ship or a recognized leader of any sort. I asked my own Visitor contact about that once, and she remained vague saying that I could talk to her about anything and she would pass it on."

President Kaitland answered, "Excellent question. I explored that once with my contact too and was told that they didn't need a specific leader. My contact did not elaborate but only explained that they have groups that specialize in different issues. I suppose that is not unlike us, except that my contact did not identify any one individual as a leader or decision-maker. As soon as we are finished here, I will call my contact and explore it further without giving away my reason. I can be confident that the conversation will be secure from eavesdroppers. On my first call from my Visitor contact, he called on my special phone that has an extra level of encryption. The Visitors apparently had no trouble with that encryption, but our experts here tell me that no one from Earth has a hope of breaking it."

After concluding the follow-up discussion, President Kaitland adjourned the meeting. She also asked Dr. Schneider to contact her with anything important that might come in overnight. Then she returned to her private office determined to talk to her contact Michael right away. She asked Eleanor to hold her calls for the next twenty minutes except for an emergency.

She entered Michael's code and after the usual few rings he answered with his friendly gentle voice, "Hello Rachel." She, like everyone else, had come to think of him as her personal close friend and had asked him to address her with her first name. A brief thought flashed through her mind—how can he always answer so quickly? How can he be alert for my call all the time? Is it possible that I really am only talking with a vast AI computer? With effort she shook the thought from her mind as Michael asked her how her day had gone so far.

She said, "It's been extremely busy today, Michael—very busy, indeed. That's one of the reasons I called you earlier than my usual after-dinner call. Something came up at a meeting this afternoon that I would like your help with. Do you have a little time now to talk?"

Michael chuckled and said, "You know I always have time for you, Rachel. I always like to talk with you. What's on your mind?"

She answered, "We are working on a problem here that looks to be serious. We have not been able to confirm yet whether the problem is real. If we conclude that it is genuine, we will want to advise you of the details. I was reminded today that we do not know which person on your ship is the contact for important, official communications intended for all Visitors. I'm sure you understand how in Earth societies, we have individuals in leadership and decision-making roles whether it's a national leader like me or the mayor of a city or the captain of a ship. Can you identify for me that person on your ship?"

Michael answered, "I'm happy to do that. You can use me for that role. I am always at your fingertips."

Rachel replied, "But Michael, perhaps you didn't understand what I meant. This would be an extremely important communication that would need to be conveyed directly to the highest-ranking individual on your ship."

Michael said, "I apologize. I realize now that we have neglected to explain some important things to you about our society. We hinted at it in one of our earliest answers. We intentionally downplayed it because we were concerned that it would make us appear even more alien to you than we must already seem. There were already more than enough reasons why humans might react to us in fear or revulsion. That's why we have avoided sending pictures of ourselves. We certainly do want our peoples to become friends, and we have been trying to reduce barriers to that."

"Yes, Michael, I can understand and appreciate that," Rachel said. "Some people here can overlook nearly any kind of difference while responding to kindness, and others have much more difficulty with that. Fear of what is strange has always been a problem for us."

"Yes, we did understand that from our studies of your history," Michael said. "We were once that way too. It makes sense for anyone who is vulnerable, and all living things are vulnerable in one way or another. Coming back to your question, as we have been speaking, I have been receiving encouragement to explain more to you. That admission alone tells you that there is something different about me unless you assumed that some of my shipmates are here with me listening to our conversation. In a sense, that is near the truth. You have heard how all of us have our machine interface pods. And we interact with our pods by using our ability to generate and sense changes in electric and magnetic fields. But we haven't explained yet how our pods give us a direct interface with the artificial intelligence system and subsystems of our ship.

"Each of us can, at will, select a private mode where we are mentally isolated as humans are, or we can choose an open mode that can be varied to connect to only one other single person or we can choose a group of any size and, in the extreme, we can choose the entire collective intelligence of the main AI system and all of our people on board. Our AI system is capable if asked of listening to and responding simultaneously to every single person on board this ship. It means that if we choose to or need to transmit vital information to everyone on board, we can do so instantly by simply thinking to do it. And we're actually pretty good at multitasking too. While we were speaking earlier, I queried the group that would have the greatest responsibility over a problem situation arising from Earth, and they encouraged me to explain all of this to you."

Rachel Kaitland took a deep breath and said, "I feel a little overwhelmed by all of this. But I do not feel frightened or repelled. It sounds like a wonderful capability. Frankly I feel envious. May I ask this—if an issue arises that would require a decision to act as a group or not to act, would all of you participate in the decision through your pod interfaces or would a designated group for that kind of issue make the decision or would your AI system itself make the decision?"

"An excellent question!" Michael said. "The answer is—it depends." Michael chuckled. "Obviously you need a little more. If we were traveling at full cruise speed in interstellar space and our systems detected a massive body in our path that would instantly destroy our ship if we collided with it, and if only a few nanoseconds were available before a course change must be made, then our AI system would make the decision in time. If the seriousness of threat is lower than that—if there is more time available to make the decision, then we use both methods you mentioned. It all depends on the consequences of the decision and how much time we have to consider the options. We have developed general rules based on experience about how these things are done, but as a general principle, we prefer that everyone be involved in important decisions. In a group our size, hundreds of thousands of lesser decisions are made every moment. We have specialists and special groups who take care of such things with the assistance of the AI system. With that explanation, I hope that you now see that you can depend on me to convey your communication to everyone on board immediately if that is your wish."

Rachel said, "Thank you so much, Michael. I'm glad you explained it to me. I'll relax about that now."

Michael answered, "Good, that's good. I did receive a small afterthought from the group I just mentioned. Might it be a good idea if you foreshadowed to me what your concern is? We can take it now as purely provisional to help us prepare if necessary, but hold in abeyance until further confirmation."

Rachel knew that her first query might provoke a follow-up question like this. She was tempted to explain everything immediately but held back. She said, "Michael, I would like to check one more thing first. I will call you back within less than an hour I hope. Thank you so much and goodbye for now."

"Goodbye, Rachel. I'll wait to hear from you."

Chapter 26

THE INCIDENT

Washington, DC, Thursday, 18 July, 4:00 p.m.

President Kaitland called Eleanor and asked her to have her chief of staff come to her study and then phone Dr. Jonathan Schneider and patch the call through to her there. In a few moments, Albert Thompson entered, and she explained quickly what she had in mind. Then her phone rang as Eleanor put Dr. Schneider through. The president switched her phone to "speaker" for Albert's benefit and said, "Hello Jonathan. I have a question for you as a psychologist. Why have we put off telling the Visitors about this potential threat? Why didn't we decide earlier this afternoon to let them know right then that they might be in danger so they could immediately begin to think about how they might defend themselves?"

Without hesitation Schneider answered, "Two reasons I would say. First, we hesitate to expose our weakness in not knowing whether the threat is genuine and when and how it might become real. That makes us feel that we will appear ineffectual and inadequate to the Visitors. So we delay while we try to confirm the threat.

"The second and more important reason I believe is that our minds are still conflicted. We see a clear moral obligation to warn them, but because we can so clearly see their power and cannot see into their minds, we still fear them at some level. We haven't had the opportunity to develop long-term, deep and genuine trust. We would not hesitate to warn truly trusted friends. And it gets even more complicated. To warn them to protect themselves from the very defense system we created exclusively to protect ourselves from them is making a leap of faith to the kind of trust reserved for old and truly proven friends—the kinds of friends we have known for years. This was never going to be easy for us."

"My God!" President Kaitland said. "That is exactly what I was feeling before I called you, but I hadn't been able to articulate it to myself that clearly. Thank you! I knew I was calling the right person." She then outlined the details of her conversation with Michael to Jonathan and Albert and asked, "Jonathan, have the Russians and Chinese already been alerted to increase security at their command installations?"

"Yes they have," he answered. "We shared what we have with them, and they are also beginning to recheck all involved personnel. Our own recheck is underway now too. For us, the Russians, and the Chinese—all of the control rooms will now have redoubled guard barriers and all will be on high alert."

"Thank you, Jonathan," she said. "I'll check with you again later this evening. Goodbye."

President Kaitland hung up and said to Albert, "I think now it will be very difficult for any attack squad to penetrate the extra defenses and high-alert status around all control rooms and their approaches. We have to hope they haven't already infiltrated some conspirators inside one of them—we're checking for that now."

Barely three minutes had gone by while she was saying this to Albert when the phone rang again. It was Jonathan Schneider.

"Madam President, a new intelligence report of highest priority came in while we were speaking a few minutes ago. It's not good news. Six other people have disappeared we believe as much as a day and a half ago. They were not people who were being watched as associates of the four suspects. The six are all Russian, and it is believed that they have been kidnapped. The crucial fact is that one of them is the key person who developed the final stage of the encrypted signaling system for the Russian EDS control rooms. The other five people are his mother, his wife, and their three children."

"No!" the president said. "No ... does this mean that—"

"Yes," Jonathan interrupted. "All of the security measures in the control rooms—all the secret codes that change frequently and the three-person cooperation required to activate the system—those are all in place to prevent unauthorized launch commands being sent from within the control room. But after a proper launch sequence process has been performed in the control room, the computer system sends a final encrypted signal to an antenna that radiates it to the orbiting missiles. Each missile has its own particular code. All of the control rooms are in secret locations, all are hardened and heavily guarded, and the cables to the antennae and the antennae themselves are also hardened and heavily protected.

"The conspirators must have known this. They have elected to bypass our infrastructure and instead create the only thing they need, the final launch signal. They now have the man who created the Russians' final launch step, and with his family held hostage, they have the means to force him to do whatever they want."

President Kaitland's mind was racing with even greater clarity and focus than usual. She said, "On the surface, such a strategy ought previously to have been thought out and made to be impossible. But do you assume there is some oversight or some issue that the conspirators think they can exploit to force the Russian expert to create some kind of overriding launch code?"

"Yes. We don't know what it is. But the Russians know all of what I just told you, and they are definitely worried. It was obvious they didn't want to share details, but from the little they told us, the kidnappers must have had the Russian expert's family already in their hands so they could force him to copy and take some critical computer files with him from his last appearance at his workplace. The Russian authorities are scouring every possibility now to find the kidnap victims and locate where the rogue control facility might be. But Asia is the largest continent, and the conspirators have undoubtedly been planning this for a long while."

"Thank you, Jonathan. This magnifies everything to an immediate state of emergency," President Kaitland said. "Have the Russians stopped responding to our questions now?"

"No, Madam President. But they are limiting responses."

"I'll check with you again soon for any updates," she said, "but now I need to do something quickly. Goodbye." She remarked to Albert, "I have just made an executive decision. I'm about to make that leap of faith."

Albert left her study, and Rachel immediately entered Michael's code into her phone. Michael answered quickly. "Hello again, Rachel."

Rachel explained the entire story to Michael. She saw no reason to hold back any detail. She concluded with, "I am so sorry, Michael, that a treacherous few have betrayed us and created this threat to you. Please extend our deep regret and sorrow to the others on your ship. Meanwhile we are doing everything we can to prevent the conspirators from carrying out their plan."

Michael replied, "Thank you, Rachel, for sharing all of this with us. Everything you told me is already in our AI system and distributed to all of the individuals who will need to initiate actions. We will have our systems on alert and will plan how we should respond. Don't worry. We do not blame you or your people for the actions of a few others. We are very grateful you have come to us this soon."

"Will you need to leave orbit and withdraw farther from Earth to be out of range of the missiles?" Rachel asked.

"We will consider all options," Michael answered, "but I don't think that a withdrawal will be necessary. We, of course, have been aware of your missiles from the beginning and have already thought this issue through carefully. We are very happy that all in our ship and those on Earth have enjoyed each other's company so well since we arrived. For a long while now we have thought of your missiles as only a formality. The fact that you have warned us now and that you and others are working hard to protect us confirms our trust in you. I assure you that we will not forget that."

They spoke a little more and said goodbye to each other. Rachel felt massive relief in knowing now that she had given them the chance to prepare. She felt certain that she had done the right thing. That night many people went with no or very little sleep while they worried and rechecked incoming intelligence reports.

President Kaitland and her advisors knew that the conspirators would realize their plot was exposed once news of the Russian expert's kidnapping with his family reached authorities. The conspirators would understand that they must take action soon before critical features of the EDS system could be changed and their plan rendered impotent. Everyone expected that something would happen within the next twenty-four hours. Everything depended on what the Russian expert who created the last launch step knew about any weaknesses the Russian system might have. No one would blame him for doing whatever he was ordered to do with the lives of his five closest loved ones at stake. He would know that any resistance from him would result in his family being tortured or killed one-by-one before his eyes. He would also realize that all of them would be murdered anyway once the conspirators had what they wanted. All he could hope for was that he and his family would die quickly. Everyone who understood the situation felt great sorrow for them.

Washington, DC, Friday, 19 July, 8:00 a.m.

The following morning, President Kaitland asked Jonathan Schneider to update everyone with the latest reports though most in the room had already wrestled with them through the night. Dr. Schneider concluded his update with, "The friends and associates of all of the presumed conspirators are being interrogated, but so far nothing new has come to light. Our agents in the area have not been able to turn up any new information now that the conspirators have disappeared. The Russians are focusing their investigations within their own agencies now and are not sharing anything with us. We have been using our satellites to listen for any hint of information that might be relevant, but so far we have nothing from that source either. At this point we are largely sidelined as bystanders."

"Dr. Schneider?" the president said. "What about a 'shutdown mode' for the Russian missiles? Surely that has already been considered, but I haven't heard a report of it."

"Yes, Madam President," he said. "I should have remembered to mention that right away. Before they quit talking with us last night, the Russians mentioned they had already tried to accomplish that very thing. They were light on detail, but apparently the operating system has a toggle setting that tells the missile to 'ignore' a launch sequence for a specified amount of time. It's a risky safety feature to have built into the system, and knowledge of how to throw the toggle must be protected at the highest level. Only six people in their program knew about this feature and how to switch the system from 'active status' to 'inactive standby.' The most competent and knowledgeable of those six was the man who was kidnapped with his family.

"Overnight the Russians tried to throw the toggle to shut down launch capability, but the missiles ignored it. Repeated efforts did nothing, and their other experts have concluded that the conspirators had outthought all of them. They must have somehow tricked or forced the kidnapped Russian expert into acknowledging the existence of the shutdown feature. They believe the kidnappers forced their hostage to disable that safety feature just after they kidnapped him and before anyone else was aware of the plot and watching for telltale radio signals to the missiles. They believe the hostage expert knew of some system weakness that the others have not yet thought of. The missiles routinely stay in stealth mode and do not report system changes at regular intervals. The Russians were only able to determine what happened after they specifically queried a system report from the missiles last night. The conspirators probably received that transmission from the missiles too and now know what the authorities are trying to do."

"Damn," President Kaitland sighed. "I don't like being sidelined to waiting and watching." She paused a moment and then said, "I'm wondering whether we might ask the Visitors to help us. It's an easy guess that the relay satellites they deployed for use with our mobile phones probably have other capabilities too. And it's likely that the Visitors have recordings of every mobile phone call made on the planet in the last three months and are currently analyzing them in their AI system."

At this point, she stopped abruptly. At the beginning of the morning's meeting she had relayed the details of her conversations with Michael the previous evening. But there was one thing she had not thought to mention. She began again, "I just thought of something I forgot to tell you from my talk last evening with my Visitor contact. My contact said they were aware of our missiles immediately when they arrived and had evaluated them. But since then with everyone getting along so well, they had been thinking of our missiles as 'only a formality.'

"There was so much else to think about when he said it that I let it go. But now it seems to me an odd choice of words. He might have meant that they were not worried about them because relations were so good. But now it also makes me wonder whether they might not see the missiles as a significant threat at all. He did say that he did not think it would be necessary for them to retreat from their Earth orbit."

She noticed Tom Beckworth wanting to speak. "Yes, General Beckworth?"

"Thank you, Madam President," he said. "Yes, it's a very interesting choice of words. It reminds me of the Renaissance uniform and handheld halberd that the Swiss ceremonial guards use on formal occasions at the Vatican. This particular halberd design comes from the Middle Ages, and, of course, the halberd and the similar pike are much older as ancient but deadly weapons of war. Thinking of these ceremonial guardsmen, it has always been one of my greatest worries that the nearly superhuman effort we put into creating our Earth Defense Shield was all expended to build a weapons system that might seem laughably quaint to the Visitors. Since their arrival with all of us able to see more of their technology, it's entirely possible that they are quite relaxed knowing that they can defend themselves. I worry now about how they might choose to do that."

"I see what you mean," President Kaitland said. "But surely the powerful Russian warheads … wait … yes, there are seven warheads on each missile. Even if the Ship's outer hull is as strongly constructed as Colonel Vanecek speculated, surely any one of the warheads would blast completely through doing severe internal damage. There would be maybe two Russian missiles close enough for an attack at any time. Think what that many warheads could do."

Beckworth responded, "Yes, I agree. We don't know what vulnerabilities their ship might have, but I'm confident the Visitors would want to avoid that. The important thing though is that the warheads first need to reach their ship. That is the main reason for my long-term worry about the effectiveness of the EDS. We used our best technology, but it has inherent limitations. Luckily, the Visitors placed themselves into their assigned orbit. That's important because we designed the orbits of our missiles to be optimal for rapid attack in that zone. We had to hope that the Visitors' ship would not be able to outmaneuver an attack coming from more than one direction.

"But making everything work was always going to be difficult. We did not have the technology to create pursuit vehicles like some of the public imagined. Everything depends on there being at least several missiles that can quickly reach the Visitors' ship wherever it might be in its orbit. All of you know how tricky orbital rendezvous maneuvers are. Increasing the velocity of one of the orbiting objects or decreasing it leads to counterintuitive changes in their relative positions. Thrust events must be precisely timed and calibrated. Each missile has an internal computer and guidance system to deal with that. By some miracle our engineers were able to fit the EDS missiles in time with vectoring thrusters to compensate for evasive maneuvers. But all the targeting steps require time. There is no way around that. We have no idea what the Visitors might be able to do in that amount of time."

The president responded, "General Beckworth, do I remember correctly that once the missiles are activated, there is no way to call them back or abort them?"

"That's right." he answered. "Once a successful launch sequence ignites the engines, there is no way to abort the attack. We did not have time to add that feature. It took everything we had just to make them work for an attack."

"I'm glad you reminded us of all this," the president said. "With your knowledge of the strategy of the missile locations and all the interacting components, you could speculate about what kind of plan the conspirators may have developed. I mean, it surely must have something to do with the relative positions of the Russian missiles and the Visitors' ship and perhaps some other constraints."

"Yes, thank you! I managed to update all the facts just before coming here. That's when everything I needed snapped into place. I was just about to raise it. We know that the conspirators need to act soon before they're discovered or before the Russians can tweak the missile controls. The Russian positions in the missile pattern were assigned every sixty degrees for six positions around the Earth in orbits ninety-nine thousand kilometers above the surface. They were grouped as two together, then a single one, then another two followed by another single and so on. So at any place where the Visitors' ship might be, there will be three Russian missiles relatively close to it.

"We can assume that the conspirators have set up their hiding place and equipment somewhere in the longitudes between central

Asia and Eastern Europe. All of the control room antennae for us, the Russians, and the Chinese are linked so that no matter where the Visitors' ship might be, an appropriate missile could receive its launch sequence. But the conspirators do not have that advantage if they have only one hidden antenna. Their antenna must be able to broadcast cleanly line of sight to missiles roughly overhead.

"So I'm guessing they have been planning a time when the Visitors' ship happens to be over central Asia at the same time as a group of two Russian missiles are trailing only a little behind and a little below the Ship. The Russian missiles with their slightly lower orbit would be catching up to the Ship. The conspirators could activate those two missiles and also the single Russian missile following sixty degrees behind. The first two could arrive on target quickly, and the third would arrive as a backup later but not as delayed as one might think from its distance. Each launch sequence has the present target location embedded within it. The launch triggers an inbuilt program in each missile's computer to calculate thrust vectors for the shortest possible time-of-flight to target.

"The important point here for this entire discussion is something I worked out just before this meeting. The optimal time window for that particular arrangement of the players will begin approximately three-and-a-half hours from now. That's shortly after sunset in that region." Others in the room familiar with the system and orbital mechanics were nodding in agreement.

President Kaitland said, "Thank you, General Beckworth for that crucial insight—this is what we needed. We should pass this information along to the Visitors right away."

General Beckworth replied, "I would be truly amazed if the Visitors had not already worked it out within minutes of your warning last evening. And I would guess that the Russians are already on it too. But I agree. We should let all of them know. I just wish we had a way to pinpoint where the conspirators are hiding. Maybe the Russians will have a breakthrough."

Washington, DC, Friday, 19 July, 12:30 p.m.

Three hours later, everyone from the morning meeting was again gathered in the Situation Room. The president had passed General Beckworth's idea on to Michael that morning. If the Visitors had already worked it out for themselves, Michael politely did not mention it and thanked President Kaitland warmly for sharing it with them. The Russians also thanked the Americans and added that they were still feverishly searching for clues to the whereabouts of the conspirators and the kidnap victims. No one knew if the conspirators would act in the next hour or two, but the odds seemed likely. The Russians were still searching for any system weakness that their kidnapped expert might have been forced to use. If they could identify it, they might be able to restore and then activate the safety feature. At best it would be difficult because the missile systems required multiple checks and confirmations before they would accept any changes. The systems were designed to be protected from alterations that could inactivate them.

The people in the Situation Room were fully engaged in monitoring all relevant satellites able to provide data from the west-central Asia region. Graphic displays showed in real time the relative positions of the Visitors' ship and the three Russian missiles in question. The radio frequencies used for the Russian launch system were being monitored continuously. Radar data from installations in Eastern Europe were also patched in. Everyone was glancing at the clock and wondering what might come next. President Kaitland had called Michael who readily agreed to being connected into the room's sound system. Michael had already informed the group that the Visitors agreed that an action could be expected soon, and they were monitoring all of their systems closely, especially the Russian launch frequencies. Everyone in the room sat silently contemplating the slowly changing digits on the face of the clock or watching the slowly shifting positions of the Ship and missiles in the radar display.

While she watched the various monitors, President Kaitland mentally reviewed the factors that contributed to the surreal and even absurd scenario they found themselves playing out. Here they were, dealing desperately with one of the most lethal weapons systems of war Earth had ever created—one designed specifically to be used against an alien invader. They were now in contact with one of those aliens doing their best to protect all of the aliens from that very same lethal weapons system. Moreover they were being assisted in their efforts by the builders of that part of the weapons system they were attempting to disable. And those builders were themselves the long-term, on-again-off-again adversaries of the US. There was more than enough irony to go around. *How did we possibly come up with this ridiculous confluence of events that led to this situation?* She knew the answer of course and unwillingly found herself thinking of the old joke about there not being intelligent life on Earth.

◆

A shattering klaxon-like alarm suddenly filled the room.

Everyone in the room jumped.

It was the alarm set up to be triggered by signals in the Russian launch code frequencies. Several things happened faster than anyone in the room could know. For the moment all they could see was that sensors had detected activity in the Russian launch frequencies somewhere in west-central Asia. A computer voice from one of the monitors announced that the two Russian missiles closest to the Ship had just been activated—their engines were firing. The main radar screen highlighted the two activated missiles and showed that they had begun to move toward the Visitors' ship. Disorientated by the deafening alarm, the people in the room were trying to regain their equilibrium while they watched the missiles slowly advancing on the Ship.

At that moment, Michael's gentle voice intervened but was barely heard. He said, "Please do not worry. We have the danger under control."

THE INCIDENT

President Kaitland shouted, "Someone turn off that damn alarm now!"

In seconds it went silent. Then she said a little too loudly, "Michael, what do you mean? We can still see the missiles advancing on your ship."

"Yes," he said. "But watch a few more seconds."

All eyes focused on the radar screen, and soon everyone saw that the missiles were beginning to veer off their course. Then, suddenly, both missiles together shot off the edge of the screen at a tremendous velocity leaving the great Visitors' ship untouched.

Michael spoke again, "The missiles' systems are shielded against ordinary jamming attempts, but we have disabled everything in them with a high-energy pulse. We mentioned before that we have tools for manipulating objects outside but near our ship. We used such a tool just now to reorient the velocity vectors of the missiles. Remember that our vessel and all the missiles are in orbit around the Earth and, therefore, are moving with Earth along its orbital path. Earth's orbital velocity around your Sun is nearly thirty kilometers per second. We erased that orbital velocity from both attacking missiles by accelerating them to thirty kilometers per second in the opposite direction of Earth's movement around your Sun. At the same time we also provided a strong impulse to accelerate both missiles onto a radial vector aimed directly into your Sun. Now both missiles will escape Earth's gravity and are falling like stones at a very high velocity directly toward the Sun being pulled even faster every second by the Sun's gravity. They will be vaporized in the Sun's outer atmosphere. We knew that our first energy pulse would not trigger the detonators for the missile warheads, but there is a small possibility that some internal components might be destabilized. In an abundance of caution, we have chosen to dispose of both missiles permanently."

For a moment, no one in the room knew how to react to Michael's words. President Kaitland felt relief as she absorbed his meaning, but at the same time unhappy questions came to mind. *Why did the Visitors let them go through so much anxiety and effort if they knew all along they could block the missiles so easily? Was the entire EDS system such a joke to them that they wanted to rub our faces in it?*

Most of the people in the room were still looking around with questioning expressions. But some were thinking of these same questions too and beginning to show puzzled frowns even while relief struggled to take hold.

President Kaitland thought to ask, "What about the third Russian missile? Have you inactivated it too?"

"No, it was never launched." Michael answered. "That missile required a separate launch sequence. Before the conspirators had time to send it, we had already destroyed their launch system with a different high-energy pulse. The two launch sequences they broadcast for the first two missiles took a third of a second to reach us. Just ninety milliseconds later our AI system identified the precise location of their launching antenna and beamed a high-energy pulse to destroy the electronic systems connected to the antenna. Unless the connected facility was deep underground, the pulse would also have destroyed any other electronic equipment there and would instantly have heated metal objects there to a high temperature. We immediately sent the antenna coordinates to the Russian authorities and now have heard that they are closing in on the location."

There were so many images and questions flying through people's minds that it was difficult to think how to respond to what Michael was telling them. President Kaitland was the first to speak with the obvious question. "Michael, why did you not tell us what you would be able to do? Why did you put us through all this worry and anxiety? I'm very surprised that you would do this to us!"

Michael's voice was subdued. It was clear he knew he had healing to do. He said softly, "We felt that we had to do it this way to have any hope of saving the lives of the kidnapped family."

That quiet statement arrested everyone's agitated emotions; they calmed and listened.

He continued, "We knew about the media campaigns intended to turn Earth's people against us, but we knew nothing about the conspiracy to attack us until you informed us. We were horrified that the lives of the kidnapped family were put at risk because of us. We tried but were not able to locate where they were being held. After what you told us last night, we searched our records and found the earlier signal that the conspirators sent to disable the missiles' safety system. Unfortunately that monitoring system in our ship does not include a close enough record of the signal's point of origin to be helpful. We dared not tell you or the Russian authorities even that much about what we could do. We realized that if we let anyone on Earth know what we could do with the missiles, then that information might well be leaked. If the conspirators found out, and we believed that would be possible, then they would know that their attack could not succeed. In that event we concluded they would murder the family and disband into hiding. We are very sorry that so many of you have suffered so much anxiety, but we felt that we must do whatever we could to save the kidnapped victims."

President Kaitland said, "You mentioned the Russian authorities are already closing in on the kidnappers. Won't the kidnapped family be carried off or harmed in the time it takes the authorities to reach their location?"

"It was our hope," Michael answered, "that in the confusion caused by the main energy pulse, the conspirators would be disoriented. But to be sure they could not harm the kidnapped family, we sent other energy beams to add confusion. One was a brilliant, intense flashing light over the whole area. Another high-energy, pulsed beam caused the building to vibrate in the infrasonic range to produce the psychoacoustic effects of fear and dread. The most important energy beam was a particular mix of frequencies that causes temporary unconsciousness, but it requires nearly a half minute of exposure to be effective. That's why we needed to induce confusion with the other beams until the kidnappers fell unconscious. It was good that you warned us yesterday evening—it gave us time to prepare."

Those in the room reeled at this new revelation of yet more of the Visitors' powers but put their questions aside for the time being. Relief was finally the dominant emotion, and all could see the logic of the Visitors' plan.

President Kaitland said, "Thank you, Michael. Now I feel that I should apologize to you for doubting you. We need to talk more of this, but first we must find out what is happening with the Russian authorities. Have you heard any more from them?"

Michael said, "Yes, I am receiving a continuous feed of updates from them as we speak. They have just arrived at the site and found everyone there still unconscious. They found the kidnapped family tied up in a locked basement room. All six are alive. We are very happy to learn this just now."

Chapter 27

AFTERMATH

Earth, 20 July to 4 August

It was inevitable that the story of the conspiracy, kidnapping, and dramatic outcome would leak. Government officials successfully maintained secrecy prior to the attack, but too many people knew the details and afterward could not contain the story within themselves. Soon the entire sequence of events was being repeated everywhere in the media and miraculously with a high degree of accuracy. Both the Visitors and the governments involved were hailed as heroes, and the Visitors now were not merely popular but were adored. The conspirators and those behind them could not have envisioned a worse outcome. The Visitors asked that the conspirators not be treated harshly—they had been misguided and could be redeemed. The Russian authorities remained politely noncommittal about that request.

The Visitors' new powers revealed by these events preoccupied almost everyone in the aftermath. During the billions of phone calls people quickly made to their contacts about the incident, questions they put to the Visitors soon coaxed from them the original reason they had not wanted to tell humans that the EDS was no threat to them. The Visitors explained that they knew the EDS was a psychologically critical support for humans when they arrived and before there had been time to establish personal friendships. They knew that in the presence of their immense ship, humans must not feel helpless and without the dignity of their own power in that first meeting. The Visitors wanted the possibility of friendship and did not want to be seen as overwhelmingly superior conquerors. They had not foreseen how hiding their power might tempt some humans to undertake the recent futile conspiracy.

Every physicist on Earth was awestruck by what the Visitors had demonstrated. They remembered from January how the Visitors had mentioned their "tools" that could manipulate objects outside their ship. But no one imagined anything like what had happened even though they had heard how the Visitors could nudge the geostationary satellites from a distance of sixty thousand kilometers to keep them on station. Many a physicist lay awake that night wondering how to penetrate the secrets of their powers—it was all so tantalizing. Some found a shred of solace in a statement Gerry had unintentionally popularized. His Visitor contact said it to him one day, and Gerry liked it so well that he shared it with a few friends. It rapidly spread.

"Just knowing that a thing is possible is a huge advantage; it takes much longer to sort through the infinity of the impossible."

No one said it would be easy though. After all, how long did it take humans to work out how to build heavier-than-air flying machines even though they had watched birds' aerial acrobatics for thousands of years? Developments happen when their time is right. A humorist in the late-twentieth century noted that humans put a man on the moon before they realized that it would be a good idea to put wheels on suitcases.

In the days after the incident, two major streams of discussion dominated the media talk shows. One was what should Earth do about the Earth Defense Shield system. There was no point keeping it any longer, and it was extremely expensive to maintain. A few, who were obviously in firm denial of recent events, lobbied for keeping it in place just in case Earth was still being deceived by a clever charade. The great majority of polls though favored a "beat swords into plowshares" approach, but no agreement had been reached on what to do with the missiles. The nuclear warheads needed to be recovered and safely stored, but what of the rest? They might be adapted into supply transporters for the Moon base, but it would be expensive to do in their present location. Another idea was to move them together and use them as a nucleus for a new, high-orbit international space station. Nothing was settled, though, and debate continued.

The second major discussion topic that appeared at that time was quite unexpected. A group of historians, anthropologists, and ethnologists had gathered and jointly concluded that no foreseeable event could better justify the beginning of a new era of human history than the arrival of the Visitors. Their press announcement proclaimed:

"Whereas the Visitors have had such a massive and profound impact on all of humanity since 1 January 2052 and whereas their arrival and influence will always exist as a pivotal transition in human history when humanity learned it was not alone in the universe, be it here proposed that the present year of 2052 be renamed as the year 1 of our New Current Era."

The group nodded to tradition by suggesting that the previous AD for *anno Domini* be changed to AN for *anno Nuntii*—the year of the message.

At any previous time such a proposal would have been dismissed with laughter. But at this time with the high popularity of the Visitors, a surprising surge of support appeared, especially among the young. A significant number of older people who had lost loved ones objected that they did not want to mark their loved one's passing as some negative number BCE—Before the Current Era. But young people with their eyes on the future didn't mind. They loved the idea of being able to say they lived during the year 1 AN. By the time intense media discussion died down, the question had settled into a majority view that such a change would probably happen in the future. It would take more time, though, for enough people to embrace the notion.

Replacing that topic was another that had been simmering all along—how could Earth persuade the Visitors to share more of their astonishing technology? Even a few of their technical powers would enable Earth to leapfrog thousands of years of painstaking effort. But the Visitors remained firm in not sharing knowledge that would be dangerous or harmful to humanity's optimal development.

This question was becoming more of an issue because the Visitors had recently begun making more references in their conversations to their need to return home soon. Thousands of callers to talk programs confirmed that their Visitor contacts routinely said that the time was not far away when they would leave, but no specific date had been mentioned. The Visitors were well aware that billions of Earth's people had grown dependent on daily conversations with them, and they realized they would need to provide a forewarning and some level of support and consolation before they said their goodbyes. The Visitors seemed to be laying the groundwork for their farewell.

Not everyone was obsessed with the Visitors' technological prowess. A significant portion of humanity was fascinated by the Visitors' description of their success in banishing war to the distant past. Callers to talk shows asked, "If we cannot handle an interstellar space drive yet, why not help us learn how to end our routine violence and wars?"

The Visitors must have known that topic would be far less well received than a wonderful new technology. There would be no arcane concepts or enigmatic equations involved to add mystery and awe to a brand new field of science with a brand new technology born of it. Even the people who had asked for the secret of ending war realized that it might only have the appeal of a lecture from one's mother to sit up straight, pay attention, and behave.

However, there were Earthlings who had concluded that the many billions of hours of personal conversations between the Visitors and

Earth's multitudes could have been a deliberate softening-up operation—a way of preparing human minds to better receive the Visitors' message of peace and harmony. If the Visitors had any reservations about how their ideas would be received, they didn't hint at them—they only hinted that something was in the offing soon.

Chapter 28

THE GIFT

Earth, Monday, 5 August

As Sunday the 4th of August became Monday the 5th at the International Date Line, the Visitors announced in a new broadcast that they had selected their departure date. They would leave Earth on Friday, the 23rd of August. They added that they would broadcast a special message to everyone beginning in twenty-four hours on Tuesday, 6 August. That brief message then began to repeat scrolling across screens for the next full day. It included a down-counting clock showing the time remaining before the special message would begin.

Eighteen minutes before midnight, Monday, 5 August at the International Date Line, the screens displaying the earlier brief message changed to display, "Please Stand By". On the hour, a new message began to scroll down the screen accompanied by the same gentle voice of all the original broadcasts. The Visitors sent their broadcast in the twenty-three languages they had most recently used.

"Dear People of Earth, we greet you on this new day with our warmest regards but also with sorrow to confirm that it will soon be time for us to leave you and begin our long journey home. We first thank you most sincerely for the hospitality you have shown us during our visit. We know that our first greeting on the first day of this year was a shock to you. We are happy that your resilience and your openness have allowed us to become your friends. In our time here we have learned much about you and have found that you and we share remarkable similarities despite having evolved on far-distant worlds. We have not learned as much about the other intelligent life-form we have encountered, and so we cannot generalize. But from our two examples, we do wonder if the evolution of intelligence tends toward a similar pattern.

"In our two cases we both exist as individuals but also as social beings. As individuals, each person maintains a necessary level of self-interest, and as a social being, each must adhere to some level of cooperation with others. We have studied the development of your own evolutionary theories and see that you, just as we, have discovered how traits for cooperation and even altruism can be established in a species at the same time that individuals compete with others for survival. The crucible of time and natural selection found those traits of intelligence and cooperation that made it possible for both of us to develop a broad systematic science.

"We are aware that many of you are disappointed that we have elected not to share our scientific advances with you. We know that the reasons we give you for not sharing may be difficult to accept. We are sorry for any slight you feel, but our longer history has shown us the wisdom of protecting you from too much too soon. In taking this position, we are not showing a lack of faith in you. We are confident that in time you will master all of the knowledge we possess and will do it organically and authentically a step at a time. As you achieve the greater powers of a deeper understanding of the universe, you will also have the opportunity to reach the understanding needed to use those powers wisely.

"In the meantime, you have the opportunity to make your present world a happier one. You can seize that opportunity to devise a better way for all to live together and better ways to care for your mother planet. We are more than happy to share with you what we have learned through our long history of living together. They are concepts that led us to what we call our 'Great Understanding.' The way we learned to live together is what has made it possible for us to be here now rather than to be an extinct civilization that destroyed itself. It was not an easy change for us to achieve, but we did. We are confident it is possible for you too.

As we describe these concepts to you, they will not come as a thunderbolt—a startling revelation. Indeed, they might seem ordinary and prosaic. But with your patience, we hope that you will give our story a hearing. Our story is about concepts for a moral structure that has actually worked for us.

"It is useful to think of a moral structure or a framework of ethics as a kind of social technology devised to make social living more harmonious while still providing for the health and happiness of the individual. As with any technology, it must be fashioned to integrate smoothly with aspects of the reality in which it functions—designed to mesh with the properties that are inherent to the beings who will use it.

"Within every Earth culture, native moral structures have already evolved. Essentially all have common features, but those similarities are often obscured under a superstructure of idiosyncratic cultural overlays. Earth's philosophers have attempted to rationalize moral structures and find a way to justify any one of them according to some universal principle agreeable to all. Normative moral theories have been advanced, but none have earned acceptance by all. We have studied your philosophers and find much intriguing thought similar to thinking from our own history. We, like you, searched for some universal basis for moral values with which all could agree.

"One philosopher from your past, David Hume, is known for an observation he made about values—the observation that became known as the *Is-Ought* problem. He concluded that logical reasoning cannot carry one from knowledge of what *is* true about the world as facts to knowledge of what one *ought* to do in the world as a moral obligation. In his view, facts about features of the world and knowledge about how one ought to behave in the world are different kinds of awareness coming from different sources. *Is*-facts come from empirical observation of the world and *Ought*-behaviors come from values which arguably are subjective to the holder of the values. Many of your philosophers examined that puzzle and looked for a way to bridge the gap between *Is* and *Ought*. They wanted to find an answer as clear to all as mathematics, but still there is disagreement over what universal principle could justify normative moral statements.

"Among your metaethical philosophers, those referred to as *cognitivists* debate with the *noncognitivists* about whether moral statements even qualify as knowable truths at all. Cognitivists say yes; they can be known. Noncognitivists say no, they do not exist as true knowledge—they are only emotions or intuitions and are necessarily subjective—what is wrong for one could be right for another. Different arguments continue while the search to bridge the *Is-Ought* gap goes on.

"Different viewpoints are considered in different ways. For example, people of faith believe a divine directive is justification enough, but different people believe in different gods or none at all. Taking another approach, a number of so-called cognitive naturalists—people who believe valid moral truths *can* be derived from observations of the natural world—have offered arguments to bridge the *Is-Ought* gap. But these, too, insofar as providing an unassailable touchstone to justify normative moral principles, have not survived scrutiny by critics who call those views a *naturalistic fallacy*. Many arguments given many names have come and gone.

"We also needed to solve this same problem. Long ago we experienced war and suffered greatly because of it. Eventually we developed a clearer way of thinking about ourselves—an insight that helped us move past violence against each other. We have noticed that you are taking steps in the same direction. A century ago your United Nations adopted the 'Universal Declaration of Human Rights' after a terrible war. After that war you also embraced the concept of 'crimes against humanity.' You have begun to discard attitudes that once excluded many people from your human family, and you now are recognizing that all humans should have the status of family. You already have long-held moral systems that contain the critical elements of empathy, compassion, and fairness. You are close to the ways of thinking that could liberate you from the old patterns that lead to violence and vengeance.

"Long ago we were socially much like you are now. We had several diverse cultures on our home planet. Each culture had a long-established moral code that had developed organically over a long period, and each culture valued its code highly. We recognized morality as a necessity that made it possible for so many of us to live together peaceably. We also noticed that the different moral codes had common elements as if they had partly converged toward each other over time or perhaps much earlier had originated from common roots. They prescribed many of the same behaviors that make living together more harmonious. However they also had differences unique to each culture that were often problematic for the other cultures. Because of those differences, there was still conflict between groups and argument over which way was best. And the fact that only a portion of each population consistently followed their own moral code's principles added greatly to the problem. In an effort to bring peace, our thinkers stripped away the differences from all of them and focused only on the basic core tenets of each code.

"Then we began to see more clearly what we needed—we needed to start with what kind of beings we were within our natural world. We certainly understood that the way the natural world deals with living creatures could not be taken as a moral guide. The natural world is not cruel; it simply does not care. It is indifferent, wasting living creatures with abandon and without a glance or a thought to their suffering or fate. The fittest survive for a short time and then are gone, replaced by others. Accident, predators, or time wipes all of them away. We needed instead to look for specific features of our nature that might be relevant to our quest.

We were not a species of solitary individuals—we were social beings. We knew that our evolutionary heritage left us with psychological traits that improved our chances of survival. Many of these survival traits were opposite in character. Selfish concern for oneself and one's offspring along with aggression and even violence to achieve advantage were traits that suited one kind of environment. On the other hand, as social beings, empathy, compassion, and willingness to cooperate improved our survival fitness in other circumstances.

"A species needs to have all of these traits at different times, but they need to be kept in balance. For a sapient species like ourselves, those traits needed to be balanced by rational thought. We saw that our traits for selfishness, power, and aggression were strong—strong enough to undermine our society and to routinely overpower our countering traits of compassion and cooperation. For us, an ethical structure and its moral code were needed to shift the balance toward empathy, fairness, and cooperation.

We began to see what was essential. We asked, what are the most basic core aspects of life that every healthy person values most highly? We came to realize that no society could be truly stable and harmonious unless those basic core values of its citizens were supported and honored for all. We came to recognize them as fundamental facts of our species that are as important as facts about our physiology.

"For example, we recognized that the most basic and central value for each individual was to continue living at any given moment in preference to dying. Yes, it is true that there are pathologies both physical and mental that can interfere with one's judgment about the value of one's own life. But for healthy people, it was an empirical fact that, at each moment, each preferred life to death.

"And the next most important value was that each person also wanted their loved ones to continue living rather than dying. We included both of these central values as supports under a moral *ought* for preserving life in preference to death. Similarly, vital needs of life such as food, water, oxygen, and a safe environment should also be included under the choice for life over death. It was apparent that preserving and honoring each person's life ought to be one of our core moral values.

"We sorted through the many facts about ourselves and settled on a fundamental group that we called our 'core values'—values that essentially all in our society agreed with. For us, choosing a moral *ought* that was supported by our preexisting common values did not require grand philosophical justification. We saw the obvious—we as intelligent social beings were ourselves existing features of our world. These core values of our species were empirical facts of our existence that were as important to our health and happiness as adequate food and water. For them, there was no gap to bridge between *is* and *ought*. They were there all along—part of our '*is* world' emerging out of our evolutionary history like our eyes, our six arms, and our brains.

"Our goal was to improve the harmony and the happiness of our lives as social beings. Yes, we did understand that by embracing that goal, we were beginning with a stated value already established. One of your noncognitivist philosophers might say we fell into a 'naturalistic fallacy' or into a contrived, subjective goal-oriented fallacy, but we would respectfully disagree. We would say that although such judgments could be meaningful in the context of such a philosopher's narrow definitions, they are dismissive of the greater argument. We would say that although the universe does not exhibit obvious purpose, we live our lives differently. We are living beings that at our best act with rational purpose or else what is the point? Our self-chosen purpose is to live happily while fulfilling our individual lives within a harmonious society. We did not disprove Hume's Is-Ought problem or ignore it; we saw it as not relevant to our quest. Our evolved core values are a fact about us and a real aspect of the real world.

"After getting this far, we noticed an analogy with mathematics; a kind of knowledge that all of our cultures could agree with. Mathematics is full of chains of logical reasoning used to create new theorems and proofs. The logic chains can be examined by anyone to verify their validity. But the logic, even if precisely correct, would be meaningless without beginning statements that everyone agreed with. In mathematics these beginning foundation statements are called axioms. They are accepted by all as self-evidently true. This is their strength. Everyone can see that they are true, and from that point onward, only the labor of logical reasoning is needed to make advances to more detailed theorems that stand firmly on the foundation of those axioms.

"Moral reasoning is similar. While it is true that the empirical 'value facts' we identified are not strictly identical with mathematical axioms such as 'A plus B equals B plus A,' the critical point is that our core values were agreed to be true in relation to ourselves—true for the kind of beings we are. If the beginning statements are agreed as being true—'value axioms' so to speak—then other more detailed and less obvious insights can be derived from them through reason. If the logical chain of reasoning is correct, then one can be confident in the conclusion reached. In developing our moral code, we began with our core values and formulated them first into a small number of moral statements that we named *The Basics*.

"We used these Basics as a foundation to support more detailed moral statements built with reason. We created our moral code and ultimately a supporting system of fair laws that were necessary for the functioning of our complex civilization. It required time to develop and time for everyone to see its worth and to understand its logic. Eventually it became the basis of our present cultural era and at last brought us an abiding peace. Our current era has endured for more than one hundred and sixty thousand of your years. We call the beginning of this period *The Great Understanding*.

"There is yet another critical element that we need to explain—namely the necessity of widespread adoption of the principles by the people. Nothing would matter unless each person agreed implicitly that they needed to freely give to all others the same consideration they hoped to receive from others. Our old moral codes suffered from this problem. Only a portion of each culture followed them consistently. For others, the temptations of immoral behavior for profit or pleasure often overpowered their conscience. In two of the old cultures, the moral code was believed to originate in a divine commandment from a deity. Immoral behavior was said to bring dreadful punishment after death, and moral behavior was said to bring wonderful rewards after death. But in life, many surrendered to the impulse of the moment forgetting the far future in the face of immediate pleasure.

"Another of the old cultures supported its moral code with extremely detailed laws bearing extremely serious penalties in the present life for the worst immoral behavior. But in this culture many risked escaping detection and punishment by the Law while surrendering to temptation.

"In all cases for all of our old cultures, fear of eventual punishment and anticipation of eventual reward were not enough. It was clear that only those people who loved 'the good' for its own sake consistently wished to be good and were able to follow a moral path more faithfully. We could see that our goal needed to be to create a society that people loved because it was good for all.

"After the formulation of The Basics, there was a long transition period—a slow departure from the old cultural ways toward the new. This period coincided with startling advances in science and technology that made possible greater physical security and prosperity. We reminded each other often to remember The Basics. The logic of The Basics was attractive and gradually became embedded in the minds of people. As that happened, more people began to better understand the source of our conflicts and see that large-scale conflict mainly arose from long-standing traditional features of injustice in our cultures. Some forms of injustice were so old that they seemed normal or even inevitable. Our technology had made basic necessities abundant enough for all—there was no valid reason why anyone should be deprived of them. The old forms of injustice that had persisted became more glaring—more obvious and impossible to defend. A change in thinking takes time and does not truly happen by being imposed. But it can change through new understanding that emerges from conversation with others or with oneself.

"We, like you, have a keen perception of unfairness. Your own scientists have shown that even some of your animal species grasp the concept of unfairness. In time, all of us saw how social injustice continued to cause chronic conflict. All of us saw how pleasant the world could be if only everyone adhered to The Basics. Finally government jurisdictions were persuaded to make changes, to make amends and to focus consistently on resolving social injustice. We as a people did understand that we would always prefer life to death, that we would always want to preserve the lives of our loved ones, that we would always prefer fairness in our dealings with each other, that we would always prefer honesty and dignity between each other. That is the core of The Basics. These understandings became a part of us.

"In hindsight, it is surprising how such long-established patterns from the old cultures could change—cultures where moral behavior was something to speak well of but which was often ignored as an inconvenience. Now moral behavior is recognized as the single most important aspect of our culture—the principles that make everything we treasure possible.

"Consider this analogy from Earth: Remember how on Earth it was once considered normal and reasonable to kidnap people and force them into lifelong slavery, to buy and sell them as property, and to give them no rights as beings like those who bought and sold them. Now such a practice is utterly abhorrent and inconceivable for any decent person. It is a deep change.

"We were able to make such deep changes too. We gave a name to the kind of world The Basics could call into being for us—we named it *The Good*. We often say to each other—remember The Basics—always love The Good.

"That is our story. We believe you could create a similar story for yourselves. From what we know of you and your existing moral codes, we believe that you possess essentially the same core values that we do. Although your philosophers will continue examining all aspects of life, we expect that you do not need either a priest or a philosopher to persuade you that you will always prefer life to death and you will always want your loved ones to live rather than die. You will not need divine or philosophical justification to convince you that murder is wrong. You will need no great universal principle to understand that if you refrain from harming others, others will have more reason to refrain from harming you. You will find your deepest core values common to all. It will take time, but it can happen with the aid of goodwill.

"As you would certainly know, there are times when moral obligations conflict—instances when the needs of one conflict with the needs of another. Sometimes these conflicts force agonizing choices to be made. When these instances occur, it is not a failure of goodwill or a failure of your basic values—it is a failure dealt by an unkind circumstance. We have found that our Basics help us resolve such conflicts as best we can.

"Please understand that we do not wish to impose our way of looking at these concepts onto you. We only tell you of them as a way of sharing our experience. If you have found our story interesting, then you might also be interested in seeing what we developed as our Basics. They are simple, and they are few. If you find merit in this approach, it will be up to you to develop Basics tailored to yourselves.

"What we have shared with you in this message is only an overview. We will send you more in the form of chronicles from our history. We will leave you now with these simple few—The Basics that emerged from our studies of ourselves."

Value the lives of others as you value your own life and the lives of those dear to you.

Treat with compassion the suffering of others as you would want them to treat you.

Honor the labor of others fairly and the benefit of their labor as their own as you would want them to honor yours.

In all things be honest save only at times to protect others.

Respect the dignity of others as you would want them to respect yours.

Reach first for empathy and last for anger.

Reach first for fairness and last for judgment.

Honor Reason as your brightest light but Reason always in the embrace of empathy and fairness.

Chapter 29

FAREWELL

Earth, Wednesday, 7 August to Friday, 23 August

In the days after the Visitors' message, Earth's people were divided in their reaction. Their opinions ranged from its being admirable and helpful to being pompous and preachy. The media discussed it widely, and numerous experts on moral theory pontificated at length. Some philosophers thought it had merit, and others dismissed it as overblown and unsophisticated. Religious leaders approved of the spirit of the Visitors' moral principles but were critical of the absence of any connection to the divine. In their view, without a link to their god, there could be no true morality. Historians and philosophers alike though were happy that the Visitors had sent their historical chronicles and looked forward to finding more substance there.

The Visitors expressed neither surprise nor disappointment in Earth's response to the history of their Great Understanding. In the days after they delivered their story, they listened to all of the media coverage, and Visitor contacts discussed the ideas at length in billions of phone calls with their Earth friends. From what those friends could determine, the Visitors only hoped that the ideas would settle into receptive minds and over time become a force for good. The Visitors were accustomed to taking a longer view.

General Tom Beckworth was one of those callers who spent much of his free time talking with his Visitor contact in those last days before their departure. He, like so many others and against the odds, had developed a deep friendship with his alien friend. Ian had patiently endured spirited questioning from Tom from the very beginning and expected no less in Tom's first call to him the day after the broadcast finished. Tom still carried the attitudes from his career as a defense analyst and expressed his skepticism that the moral structure extolled by the Visitors had the power to end war.

Tom said, "I realize you explained that the change from your Great Understanding took place slowly over a long period, but I still can't see how it alone could ever end war. Don't you still need weapons and an army to protect yourselves against a surprise attack? Without those defenses, aren't you just an inviting target?"

"Tell me, Tom," Ian replied, "when you go to visit your best friend, Joe Garcia, are you always careful to carry a loaded pistol in case Joe suddenly attacks you?"

"Of course not. That's ridiculous! Are you trying to tell me that it's all sweetness and love across all the many trillions of your people in your home region? To me that sounds ridiculous too," Tom answered.

"Well, sweetness and love might be a touch overstating it, but certainly there is a high level of trust and deep friendship across our people," Ian said. "It's all tied to the ongoing maintenance and practice of fairness to all and honoring the dignity of all. That is the commitment of each person, and it is strengthened through our long standing traditions and institutions. Each person sees how important it is. Of course it's also true that our medical science has overcome the kinds of mental pathologies that still exist here. You would still need to deal with those issues with compassion as we once needed to do.

"And beyond that, our longer lifespans help each of us gain perspective. We still do have room in our society to enjoy the many eccentricities of individual personalities. That's part of the charm of our social lives. But at a deep level, we trust each other to love The Good."

FAREWELL

Tom said, "I would love to think that we could achieve the same situation here someday. It's hard for me to imagine, but it would be wonderful."

"I'm glad to hear you say that, my dear friend. It is all part of loving The Good. Keep thinking that way," Ian replied.

As the day of departure drew nearer, Tom had many more talks with Ian. Tom always strived to be strong and self-sufficient, especially since he lost his wife. Only with his trusted friend Joe did he expose any of his vulnerability, and it was a mystery to him how he reached that same level of trust so quickly with Ian. He gradually accepted that it happened because Ian fulfilled a long-held need he had for a constant and intimate friend. He had learned much from Ian and had rediscovered emotional depths that had long been suppressed. On one of their talks during the last days, Tom was feeling an intense sadness in knowing that Ian would soon be gone and gone in a way as permanent for him and as final as death.

Tom abruptly brought up his feelings in one of their last conversations. "Ian, we have talked of this before. I know that you know it will be hard for me to say goodbye to you and not be able to talk with you anymore. I want you to know how much I have treasured you these last months and how much I will miss you."

"I will miss you too, dear friend," Ian said. "I have cherished your unfaltering questioning spirit so much. It is an unhappy fact that our destination is so far away and that we will not be able to talk any more. I have always looked forward to our conversations, and I too am distressed to say goodbye. I'm glad you brought this up, because I have been worried about you. Would you mind a small bit of advice from a friend who has lived a very long time?"

"You have already given me so much good advice. I'm always glad to have more," Tom answered.

Ian said, "I'll keep it simple. Think of how quickly our friendship grew. I know it surprised you. I believe it happened that quickly because we talked often, and after a while you did not feel threatened by me.

You were able to open up and share your real feelings with me. You were eventually able to trust me as a friend. I'm confident you can do that again with someone else if you allow yourself to reach out. It's worth the risk of failure. It's been a long time since you lost your wife. You are still a young man. From what you have told me of her, I'm certain that she would want you to be happier than you have been."

"Yes, I do believe she would want that," Tom said softly. "But I wouldn't know where to start."

"Dear Tom, I think with a little reflection you would know. Think of all those people you know well, people you have come to both respect and like. Think of those people with whom you have shared important moments of your life and have learned to trust. Surely some of those people you have shared stressful times with since January come to mind. Intense shared experiences create a bond. Those people are out there, and you will have much in common with some of them. If you reach out to one of them, you might be surprised at how welcoming the response could be. You don't know what the future holds."

"Well ... yes, there is one person I have thought of that way often—quite often actually. But I would not have dared to think of reaching out to her before. Maybe I could take a chance."

"Yes, Tom," Ian said. "I think that is an excellent idea."

❖

When President Kaitland was younger, many interested young men found her attractive and not the least for her intelligence and her balanced confidence. But she had kept them at arm's length; her ambition and her focus on her studies and career consumed all of her time. Recently she sometimes regretted that she had not made room for someone else in her life. With the Visitors' departure imminent, she felt deep sadness over the need soon to say goodbye to Michael. She had very little uncommitted time, but because Michael did not need to sleep, she talked with him late at night when she finally put aside her work to prepare for bed. He had been a huge comfort to her and had supported her through the many stresses she faced over the last months since they met. She dreaded being without him and knew his absence would leave an aching void even though she had long been accustomed to living alone.

She knew that after the Visitors left for home, the usual tensions in the world would likely reappear. She knew she would need to focus on her campaign for reelection—there were goals she still wanted to accomplish. The past months had surely been the most unusual time in a presidential election year in American history. There had been almost no campaigning from either of the major political parties. The public's perception of her handling of the Visitors' presence was extremely favorable, and she was certain to be nominated for a second term. She knew that after the Visitors left, her party's advisors would fall on her like a tsunami, and the campaign would begin in earnest. She was dreading that chaos, and it was all the more reason she would feel bereft when Michael was gone.

In one of her last conversations with Michael, Rachel told him of her sadness about his leaving. She said, "Michael, it's true isn't it that there is no way we can continue to communicate once you are on your way home?"

Michael answered, "Yes, Rachel. I'm sorry, but the laws of physics rule us all. Once we are fully underway, time and distance will become overwhelming. Soon we will be light-months away from you and then light-years away. And if we accepted the long wait times, signal transmission and reception would be difficult and undependable even if you had very powerful equipment here. I wish it were possible for us to continue to talk. I have so enjoyed my conversations with you."

"Thank you, Michael," Rachel said with tears forming in her eyes. "It hardly seems possible to me that you have become so dear to me so quickly. I will miss you so very much."

"I will miss you too, dear Rachel. I will always remember you. Our time visiting Earth and growing to know its people has been something none of us expected. Our experience here will be a great treasure for all the rest of our lives. And my experience in knowing you is something I will always cherish."

"Oh, and for me too, dear Michael," she said. Her voice was beginning to break, and words adequate to her feelings were not coming to her. Tears blurred her eyes.

"Thank you," Michael said. "I feel the same—there is a profound anguish that comes with such parting. I believe our friendship was able to grow so quickly because we weren't afraid of each other. We talked often and were able to share and accept our deepest feelings. There is something about that I wanted to say to you. Would you mind if I gave you a small bit of advice learned in my long life?"

"Please do. I have loved all the advice you have given me."

Michael said, "You and I have been able to communicate deeply as voices across space, and it has been wonderful. But you have the opportunity to build a new friendship with another person whom you can see and touch as well as hear. You are still a young woman with much time ahead of you. If you made time for another person in your life like you made for me, you might find another true friend."

"I know that you are right, Michael, and I would love that," Rachel answered. "I have often thought that. But it would be difficult for me now because of my position. It would be so difficult for me to know whom to trust like I trust you."

"Perhaps not so difficult as you fear, dear Rachel. Think of those people you already know whom you have learned to trust through arduous times. You have had many stressful moments through the last seven months, and you have worked with many good people who have helped you. There is likely to be one among them who could be that true friend for you. Intense shared experiences can be the truest way to gauge one another."

"Yes," she answered. "And yes ... there is one like that. I have thought of him more than a few times in quiet moments. He was difficult in the beginning when I first met him. But in the last several months he seems changed—he's more open, much warmer, much more approachable."

Michael said, "Please think about it more, dear Rachel. You might let him know that you have noticed him. You might reach out to him. I think you will be glad if you do. One never knows what the future holds."

◆

In those last days, Earth's own phone relay systems would never have been able to cope with the demand of the billions of simultaneous calls between the Ship and Earth. Fortunately the Visitors' system appeared to have no limits. At no other time in human history had so many people felt so overwhelmed at the same moment by such intense sadness of farewell. No one would have predicted four months earlier that in such a short time human beings and alien creatures that resembled nothing familiar could have become dearest family.

A common theme in those many last calls was a gentle suggestion by the Visitors to each of their friends to look to other Earth people for loving friendships and to remember how those friendships can deepen when trust and understanding are freely offered. It was certain that Earth's people had learned much from the Visitors, and there was good reason to hope that what had been learned would be remembered.

The four friends too felt the deep sense of loss. Sandra, Gerry, Ellen, and Jim—all shed tears during their final calls. All were heartfelt conversations pledging never to forget each other. It was the same with the billions of Earth's other people who had come to love their Visitor friends. Intensely emotional expressions of gratitude and love filled those last heart-to-heart talks—often heartrending last words and goodbyes from all those who had grown to depend on the trusted companionship of their dear gentle friends.

At last, as the clock marched into that last sad hour, the Visitors explained that it was time for them to leave. All who had been in conversation with the Visitors knew they must finally let go. At the appointed time, after the last lingering goodbyes, the Visitors shut down their relay system. Countless humans suddenly felt desolation as they looked at their now silent telephones that would never again convey their friends' gentle voices to their ears. Telescopes watching from Earth recorded how the Visitors' seven large relay antenna satellites began to move off their normal orbital path and toward the Visitors' ship as if by magic. Within six hours all of them had been brought back on board.

In the following hours, the familiar shining orb began to shrink as it slowly spiraled farther away from Earth. Perhaps as another way of saying goodbye, the great vessel resumed its full reflective brightness and shone gloriously in the sky. The Visitors had explained that they must tease themselves out of their close orbit very gently and would not dare engage their main drive engines until they were much more distant. With every capable telescope on Earth trained on the departing vessel, it spiraled farther out and gradually arced out of the planetary plane into the southern Celestial Hemisphere. Telescopes and radar facilities in Australia, South America, and South Africa again had the best view. When their instruments showed the vessel to be approximately fifty-four million kilometers south of the planetary plane, it rapidly shrank to invisibility over the course of a minute.

During those last hours while the departing vessel slowly arced outward from Earth and before it finally disappeared from view, a last message was broadcast from it that played over and over again until the Ship vanished.

"Dear Friends, we have come to know you and love you. Thank you for your hospitality during our visit and for the love you have returned to us. Perhaps some of us now alive might visit you again someday, but the universe is vast—even more so than you now understand. You will understand in the future.

Great possibilities lie ahead of you. Be confident of yourselves. And be confident of our friendship. We will not forget you. We as peoples will see each other again someday. We as peoples will see our friendship grow together into the future. Please always remember your Basics. Please always remember to love The Good. Farewell Dear Ones—farewell."

It was night in Tucson. From a small receiver, Jim and Ellen listened to the recorded gentle voice speaking these last words as they stood outside looking up at the stars. Being in Earth's northern hemisphere, they knew they could not see where the Visitors' ship had last been visible, but they were with them in their imaginations. They missed the Visitors terribly, and although they had tears in their eyes from the farewell, they also felt happier and more hopeful for the world than they had in years. They stood looking into the immensity of the night, silently thinking of the Visitors now speeding away from Earth toward their home so far away.

Since childhood, Jim had been fascinated by the night sky. Later as an adult, he noticed how people loved to be able to look out to a far distance—to see out over a great canyon or over a wide valley flanked by mountains on the far horizon. For many, a view over the great divide between land and sea, one of Earth's greatest natural boundaries, was even more admired. And yet somehow overlooked because it is always there, is the greatest natural boundary of all with the grandest view. Merely looking upward presents the greatest possible spectacle—the splendor of the vast starry universe.

Just then with the two of them together under the stars, the sky seemed to Jim even more magical than usual as he felt transcendent horizons calling to him. Now as they stood in night's expansive embrace, he sensed again the awe and mystery he felt as a child when imagining all that might be. The feeling ... oh the feeling ... an unleashed tide of the numinous lifting him up and sweeping him toward something wonderful.

He looked at Ellen and said, "You know, it's been ages since we talked about this. And I know we might be a little past our prime ... But there are so many children in the world who need loving parents. What do you think?"

Ellen looked up at him with tears still in her eyes, embraced him hard, and smiled up at the starry night above.

Jim too looked up at the stars with shining eyes. He thought back to the past—to how in his melancholy night thoughts he raged at the blank-eyed indifference and brutality of the universe. Now in the afterglow of the Visitors' last words, he wondered whether he was seeing a new side. If there is a plan—a counter-balance to the brute forces of natural law—maybe it lies in the agency of beings formed from those forces, beings like the Visitors and himself. We have the power of choice. There is an indifferent side and a compassionate side—we can choose. He smiled and thought about their future and what it might hold. Earth once had its Renaissance, and it once had its Enlightenment. *Maybe a Great Understanding is waiting for us too.*

Then swiftly without effort, that old half-forgotten epigraph from so long ago came back fully to his mind:

Dark is the night of our despair when the gods remain silent.
Bright is the morning of our hope when we wake to each other.

EPILOGUE

After the Visitors vanished from Earth's sight, they did not accelerate to their full cruising speed while yet within the Oort Cloud that surrounds the Solar System. At about forty-eight thousand AUs from the Sun where the Oort Cloud thinned significantly, the colossal ship slowed and paused long enough to leave something behind before it accelerated to speed homeward. The Visitors left a great long-lived beacon that would collect signals escaping from Earth and relay them to the Visitors' home region on special frequencies controlled by the beacon's quantum computer. The computer camouflaged the information so cleverly that the encrypted signals looked like random background noise emanating from stars. The Visitors released the beacon into a solar orbit highly oblique to the planetary plane where at that distance it would almost certainly never be noticed.

Although the beacon was in full stealth mode and reflected virtually no light or radar frequencies, it was not there as a spy. The Visitors felt it would be better if Earth did not know it was being monitored. They only hoped to learn of some of the events on Earth into the future even if the news would be more than 240 Earth years old by the time it reached them.

But the beacon had another purpose as well. In the time of The Old Ones, the Visitors considered what dangers might exist in the far reaches beyond them. They arrived at their own "Dark Forest Hypothesis" and wisely took steps to conceal their presence from possible distant listeners. When their technology was advanced enough, they set distant warning beacons in their many travels. Earth's beacon now was one of them too—a sentry alert to other signals from space that might indicate unknown travelers. The Visitors had already made plans for protecting Earth from its own signal bubble as they had done for themselves long ago. Danger could be lurking anywhere, but the Visitors were confident Earth was safe for the time being. They would carry out their plan later when Earth would be ready for the necessary technology. Meanwhile, the immensity of the galaxy itself provided a kind of protective shroud. And the Visitors' many, scattered beacons in their part of the galaxy would provide advance warning of possible danger.

The peace-loving Visitors were prepared, if need be, to protect themselves and those they love if survival requires it of them. One never knows what the future holds. The Visitors take a long view of things, and it might be that someday Earth would need their help. They would not be able to come quickly, but come they would.

ABOUT THE AUTHOR

David Loschke grew up in the central plains of the United States but feels more at home in the western US with mountain ranges in view. He has spent his career as a scientist in research and teaching both in universities and in government. A long sojourn in Australia left indelible memories of a sunny continent and a friendly people. And the night sky there is inspiration enough for stories from the stars.